MINECRAFT™

First published in Great Britain 2022 by Expanse
An imprint of HarperCollins*Publishers*
1 London Bridge Street, London SE1 9GF
www.farshore.co.uk

HarperCollins*Publishers*
1st Floor, Watermarque Building, Ringsend Road
Dublin 4, Ireland

Written by Tom Stone

ISBN 978 0 0085 2704 4
Printed in Latvia
001

This book is produced from independently certified FSC™ paper
to ensure responsible forest management.

For more information visit: www.harpercollins.co.uk/green

MINECRAFT™

MASTER BUILDS

A Showcase of Breathtaking Creations

BY TOM STONE

Contents

—

Unless otherwise stated, all builds use Minecraft's standard texture pack.

Foreword

—

One of the most amazing things about Minecraft is how we can all share
our creations, be inspired by each other, and learn.

I never considered myself a builder, not like those other people – like
the people who build the kind of creations you will find in this book. I
felt like they had something special that I was missing. I've spent many
hours watching players build amazing things and looking at the details of
their creations to figure out just how they could make something look so
good. After all, we all use the same blocks, so in every creation there is an
opportunity to discover a new style or technique.

I kept building, taking on larger projects and making videos about them,
even joining the Minecraft development team. Despite this, there was
never a single point where I felt like I had become a builder – there was
always so much left to learn! Over time I've come to realize that there is no
bar to clear. We can all build and with the boundless imagination of the
Minecraft community there is always something new to be inspired by.

Whether you are looking for amazing creations to marvel at purely
for their aesthetics or inspiration for the next project in your world,
I am sure you will find something in this book. With every player's
unique background there is always a new perspective leading to another
amazing project.

We are all builders.

Mikael "slicedlime" Hedberg
Minecraft tech lead and content creator

CHAPTER ONE

Aman, The Immortal Lands

By Iskillia

I skillia's main piece of advice to builders is, "You have to finish it," and he believes it's the secret to his success. "I really take the time but I also finish the things I begin. That's something I force myself to do. You have to finish it!" says the French builder, talented renderer, and architect of these Immortal Lands.

After eighteen months of painstaking crafting, Iskillia proved his word by finishing the vast wonder Aman, the sum of many eclectic influences. Minas Tirith, the imposing city from Peter Jackson's *Lord of the Rings* movies is a staggering feat of production design that clearly influenced the way Aman's statue and citadel manage to look as intimidating as they do beautiful. The vegetation is inspired by the gorgeous green vineyards of the south of France, which Iskillia is lucky enough to live near. The influence of the slightly soggy (but magnificent) city of Atlantis can be seen in the use of aquatic colors and grand scale, while the eighth-century Round City of Baghdad, with its unique circular design, inspired the shape of the citadel.

Considering the wide variety of influences – not to mention all the other games, architects, fellow Minecraft builders, and much more that inspired Iskillia – it's amazing that his city of Aman (the name is another Tolkien nod: a reference to the continent to the west of Middle Earth) isn't an incoherent mess. Instead, its disparate parts flow flawlessly into a whole. From the red spires – designed to resemble flames – that twist around its landscapes to the enormous white-and-gold statue, worryingly close to slamming its horned head against the 256-block height limit of version 1.12.

"When I found Minecraft, I was like 'OK, this is incredible.' I was always a creative person ... I liked the game because of the possibilities, the freedom that it offered."

Iskillia discovered the game nine years ago when he was still in high school. "When I found Minecraft, I was like 'OK, this is incredible.' I was always a creative person. I always played with things." The imaginative young builder loved to spend his time creating something new, whether an environment or a universe. So, even before Minecraft, he desired to build something big. He just needed to find the right tool to let him realize his ambitions. "I liked the game because of the possibilities, the freedom that it offered."

● **Top** | Iskillia's circular central sculpture was inspired by eighth-century Baghdad.

● **Bottom Left** | The lush green of the vegetation and the flaming red spires are a dramatic backdrop for the pale buildings.

● **Bottom Right** | The green serpentine bridge borrows its design from mythical creatures like a sea dragon or the Loch Ness Monster.

Golden God

You'll find this striking statue towering over the sanctuary part of the city. Iskillia was inspired by Egyptian gods, which often have an animal head and a human body, but he crafted an imaginary deity to make the connection less explicit. Though it looks like a stone statue, it's actually mainly made from birch wood blocks, which Iskillia realized would have a "sandstone or limestone" appearance at this scale. "The goal was to resemble an old stone monument, like the pyramids of Giza."

As for that outfit, Iskillia turned to the ancient divinities from Greece and Egypt – who weren't known for wearing a lot of clothes – for his fashion research. "I also added the flags in the back because it was interesting in terms of design shapes and composition."

After spending some time playing with friends, Iskillia entered the online Minecraft community, joining several French build teams. "I ended up in NewHeaven, where I'm a manager now." Looking at NewHeaven's portfolio, the grand cities and imaginative statues have a familiar flavor. Iskillia is also well-known in the Minecraft community for his rendering skills, a service he's offered to many builders. These rendering skills would be vital for getting an overview of his own city midway through its construction, but that's jumping ahead of the story.

Before starting the Immortal Lands, Iskillia did some initial planning: doodling designs in the margins of his school papers. But he mainly preferred to dive straight into building and develop a plan on the fly. "It's a very agile method. It changed all the time. I'm building and building and building and at the end I have something. Something that was absolutely out of control at the beginning!"

Over time a structure did develop, one Iskillia breaks down into three stages. "There's the beginning, when you're just creating things and trying to find a way to assemble them together. The second part is where you're planning and you choose a global organization of the map, the structures, etc. The final part is when you know what you want. Now you're all in on the details."

"It's a very agile method. It changed all the time. I'm building and building and building and at the end I have something. Something that was absolutely out of control at the beginning!"

This final stage was the most challenging. "You can't copy-and-paste things," believes Iskillia. "So you're forced to craft things that are very similar-but-different all over your map. I think it's the hardest part in art in general!" Look closely and you'll see that, remarkably, a lot of the similar-looking structures, particularly the many houses, do have subtle differences. It's hard to argue with how much character it gives his city, but it's a level of commitment that instills a need to lie down just from thinking about it.

Getting the minor details exactly right was tough and time-consuming, but seeing the bigger picture of Aman presented its own challenges. "When you are in-game you have a limited view and a limited view distance." It's an obstacle that naturally becomes a bigger stumbling block as your build becomes more vast. "That can be a problem. When you're working for a year and a half on the map, you get to know it really well. You can move around it easily and really know where everything is. But it can be difficult to see the 'globality' of the map."

Red Sails Under Red Skies

Loosely based on Venetian ships, this red-sailed beauty of a boat is a smart way to introduce a contrasting color to the wider build without clashing with the aesthetic of the existing architecture. Plus, Iskillia's typically absurd eye for detail stops the ship from feeling like mere window dressing and actually makes it part of this city because he's not just built the ship, he's built the ports and included the goods, ready for trading. Hay bales and pumpkins anyone?

Every Building Unique

Iskillia used birch wood for the structures of the small buildings (well, small by this build's standards). To contrast with the paleness of birch blocks, he used spruce wood for the borders and prismarine for the roofs. He felt that the cold blue/green prismarine was a color that would work nicely with the warmer colors from the wood. Wonderfully cozy!

● **Above** | The Immortal Lands in their entirety, showing the varied shape language that Iskillia utilized.

● **Right** | Iskillia thought out the balance of color in his build, ensuring each element stood out.

What Iskillia needed was a huge amount of oversight to get a good idea of how his city was progressing. That's where his aforementioned rendering skills came in. Rendering isn't just the fine art of making Minecraft look absurdly realistic. It can also let you see the full scope of your creation at last. "The first time you see the whole map in one frame," explains Iskillia. So, halfway into his eighteen-month project, he made a render that let him do just that. How did it feel to finally see nine months of his hard work in its entirety? "Whoa." Well put.

With its towering statues and intricate citadels, it's easy to look upon The Immortal Lands and imagine a real city with a rich and detailed history. In fact, many of the choices Iskillia made have turned the city into something of a record of his Minecraft skills as they evolved over the years. "I took an old build of mine. A very old build from maybe two or three years before starting this project. It was in the style and atmosphere of this map and it had interesting ideas, so I decided to add it to the map, even though it was not at the peak of what I can do."

It may seem like a strange choice to place your earlier, less-polished builds besides your best work but Iskillia is adamant that hiding your earlier creations is actually a mistake that too many builders make. "Don't judge your past self," he advises. "Many artists have this problem. They are getting better at their skill and so they judge their past work. When you're making a map over eighteen months, if you're judging your past work all the time, you won't be able to continue. You will always destroy what you've already made or move on to another project."

Iskillia's less self-destructive approach is essentially what real-world cities are – a hodgepodge of loosely connected designs and ideas, crafted during different times by builders who change and evolve the longer they work within their cities. "Absolutely. In a civilization, there aren't only beautiful buildings. There are ugly things! I think those errors, if we can even call them errors, make the map richer and more interesting." It would be hard to call anything on Iskillia's map ugly. Instead, it's much easier to admire the sheer scale of his city and enjoy discovering more delightful details upon every view.

CHAPTER TWO

A Duck Build

By SadicalMC

A Duck Build is a celebration of the poshest, most pompous flock of ducks ever. These wealthy waterfowl boast monocles, top hats, mustaches on the brims of their beaks, and even elaborate castles protruding from their bright yellow backs (because of course they're landlords). But just because these avian gents have a stiff upper beak doesn't mean they're not interesting. Perhaps, as its architect SadicalMC believes, that's because "you can just make anything funny because of a duck."

For this particular build, SadicalMC's plan was, "a line of ducks in a pond with castles on their backs, wearing top hats," which is a tough elevator pitch to beat. It certainly didn't hurt that his chosen subject was one he felt strongly about. "I love ducks. They're just so simple and so cute. I've met some people through the building community that were big duck advocates. When I was a Build Battle player on Hypixel [a popular Minecraft server] for a bit, I used to just build ducks out of everything. Ducks everywhere!" Good luck to anyone thinking of a better Minecraft building philosophy than "ducks everywhere!"

Despite his experience of duck-building, it's always useful to go back to references, so for this build, SadicalMC researched ducks online. "I did look up a reference image for rubber ducks, just so I could get an idea of how big the head should usually be. Rubber duck heads are pretty massive compared to real ducks." Handy, as that gave him room to add fancy top hats and more castle spires to their heads. His ducks manage to look a lot fluffier than the ones you usually find in a bathtub – a deliberate choice to add some surprisingly realistic details to a very fantastical build. "I also looked at pictures of real ducks. I like making my builds realistic, but also with a lot of fantasy and mythical aspects to them. I like combining color, plants, realism, and fantasy." And getting those realistic details right helps the fantastical elements shine all the brighter. The plant life, which so easily could have been an afterthought considering it's not the main focus of the build, has clearly been crafted with great care to look realistic. It's a testament to SadicalMC that the lifelike elements of the build don't seem completely out of place alongside top-hatted ducks with castles on their backs.

Right | An unusual muse, but one with a wide color palette to choose from. SadicalMC chose blocks with a bright contrast to the ducks' yellow "feathers."

● **Left Page** | These feathered friends were created using a mix of real-life and toy references, giving them giant, round heads and streamlined bodies.

● **Above** | SadicalMC had a lot of fun decorating the pond with myriad plant life plus some fantastical lanterns.

● **Left** | A closer look at the intricate towers that are artfully balanced on the ducks' backs and beaks.

You don't become the ruler of the duck pond overnight, and SadicalMC has been honing his particular set of skills for years. "I got into Minecraft because I was introduced to it by a couple of friends when I was in elementary school. It was super fun." It was the advanced technical possibilities that redstone offered that initially enticed him, such as being able to build an interactive game of tic-tac-toe. Returning to the game as a teenager, he discovered he now much preferred the building aspect of Minecraft. "As a builder now, I'm like, 'That redstone stuff is overwhelming. I don't like it!'"

Seeing the great creations the Minecraft community were constantly sharing inspired SadicalMC to try some builds himself, although he claims not to have been an overnight success: "I failed miserably." Luckily SadicalMC stuck with Minecraft, gradually improving his building skills. He credits a lot of this to seeing the relentless stream of impressive builds coming from the community. "The more I got into it, the more inspiration I saw. It just inspired me to be better as a builder and to appease my creativity."

01 The semi-transparency of yellow glass – especially when zoomed out – gives these ducks a slightly downy look. He also used white wool, white concrete powder, sponge, and gold.

02 "I chose this color palette for the castle because I really liked the yellowish-white with any kind of red or orange-toned shade." Making those castles the same color as their beaks also implies that those spires grew out of the ducks organically!

03 SadicalMC may have an eye for detail, but that doesn't mean he's willing to waste effort on unnecessary crafting. "I was debating whether to add the little feet to them, but they were going to be covered in water anyway so it wasn't worth it."

04 "I absolutely love building plants and flowers. I used a variety of green blocks." The range of blocks used makes the plant life look more alive than if he'd just used a single, solid block of color.

05 Which is SadicalMC's preferred bird? "I think the smallest one is my absolute favorite."

Recently his creativity has been focused on perfecting intricate elements. "I'm a big advocate for texturing and very fine details. That's been a trend with Minecraft lately. A lot of people are doing really small, simple builds but with fine detail, where they can manage to cram a five-by-five with the most insane details ever." It's a design philosophy he's clearly carried over to his own work. By focusing on such a deceptively small scene of four birds on a pond, he was able to go all-in on making every aspect as hyper-detailed as possible. Not a mustache hair out of place.

But let's get down to the most important question. Do these ducks have names? "No. Oh God, I should have named them though!" he laments. Luckily, this is far from his first duck build. In fact, it's not even his first build to feature fancy ducks in top hats. The previous trio were named Sir Quackles, Sir Quackles the Second, and Sir Quackles Junior. At least the builds are more inspiring than his naming conventions.

"This build was a great opportunity for me to combine my three favorite things: ducks, castles, and plants. That was the whole point of the project." Crafting a unique build around the three subjects that bring you the most joy and then letting your imagination run wild is a dream for creators around the world. Even these pompous pond-dwellers should tip their hats to that, although they may risk knocking a spire or two off.

CHAPTER THREE

Aqua Princess

By Varuna

Sunken chests. Bountiful shipwrecks. Mysterious monuments. Minecraft's oceans offer tons of treasures and wonders for anyone willing to dive in. True, they also offer sea monsters that shoot lasers from their eyes and trident-wielding underwater zombies, but there's probably just enough wonder down there to make submerging yourself worth it. If you're still not convinced, perhaps Aqua Princess from professional Minecraft build team Varuna is enough to entice you underwater.

The aquatic build is covered in clever details to give the impression of ruins that have been under the sea for a long, long time. Details like the plant life that's slowly overgrown both the princess monument and the castle she holds; the shoals of fish swimming around like they own the place; and the hammerhead shark that makes exploring this royal ruin immensely dangerous. Deep-sea builders BlackKIng89, Fiksat, ColoradoMcoil, and RenQeN, along with Varuna founder Thomas Sulikowski, have clearly sunk the time in.

"I grew up playing Minecraft. Every day after school, my friends and I would play for a few hours and that's what really grew my love for the game." Thomas soon started playing online, where he learned how to build even better from "other people who liked placing blocks creatively." Born and raised in Boston, Massachusetts, he started out by recreating buildings closer to home (quite literally – one of the first builds was his own house), including famous landmarks like Boston City Hall. He saw early success selling these builds, but more and more offers were coming in to use his talents on client commissions. "Once I started getting a lot of orders, that's when I really decided to file for Varuna to become a company." Five years later, Varuna has dabbled in circuses, music festivals, villain bases, spaceships, and some truly gargantuan city builds. "Those huge cities are pretty difficult and take a long time," says Thomas, explaining that a team will spend three or four months getting them right.

Aqua Princess started out as one of Varuna's smaller builds. The client had a loose idea of what he wanted, plus a helpful reference image of a statue. "This woman underwater with her arm extended, and I think she was holding the castle." That may have been what the client originally wanted, but Varuna soon countered with more ambitious plans. "We thought it would be even cooler if we found a way to wrap the castle around the woman, so it was all more unified.

● **Above** | This build is one that must be seen from several angles. This view displays how much higher the tower is than the princess.

"Once I started getting a lot of orders, that's when I really decided to file for Varuna to become a company."

That's when we added this long staircase going up to the castle with towers leading up to it that then wrap around her left shoulder."

A large part of Thomas's job is communicating with the clients. "Normally, when we start a commission, clients will be really excited and so they'll be asking us for screenshots. I'll send them, but they're just a bunch of crazy lines drawn in Minecraft of where every single house is going to go and outlines for where the streets are going to go. They normally look at them and go 'Oh, this doesn't look nice at all!'" Varuna also checks in with the commissioner of the build "at least once a week" to show how their build is progressing. "We'll make sure to send them the blueprint of the first house we create."

Top Left | The castle atop the princess's outstretched hand.

Bottom Left | The build team created a selection of different sealife, including clownfish and turtles, to populate the ocean around the castle.

Top Right | Thomas and his team worked meticulously to create an organic-looking human shape using the blocks.

Bottom Right | A gigantic blue tang peeks through the windows of the tower.

If the client doesn't like the plan, they can let Varuna know "before we basically recreate the house in three thousand different ways." A weekly check-in or potentially demolishing three thousand houses? Not hard to see why Thomas prefers the former.

As the team became more involved in the build, the initial concept evolved. "Originally there was just one big fish around the castle, but we added all these schools of fish." The builders researched images of lots of different fish "just to make sure they got the right dimensions" and refused to settle for just one species of fish, to make the build a wider, more varied reflection of ocean life. Smart architectural additions and bonus fish are nice extras, but they both pale in comparison to Varuna's biggest idea. "In the back, behind the statue, there's actually a shark." It's a giant hammerhead, in fact, that wasn't in the original reference image. Meticulously crafting a massive shark and then tucking it behind the princess in your main render? Well, it makes for one heck of a surprise when you enjoy the build from other angles. Another alteration they made for the statue was bad news for anyone living in the castle who doesn't want giant glowing eyes staring at them 24/7. "You're watched all the time. No break!" The decision to give the princess such eerily bright peepers was made because "that was the easiest way we could bring out the details in her eyes."

In a build full of organic life, ancient monuments, and a stealthy shark, the toughest element for Varuna to get right was a convincing haircut. "It's a very organic shape, a human being, and in a blocky world, it's hard to mimic." As if that wasn't challenging enough, the team also had to contend with Minecraft's unique interpretation of how gravity works. "The woman's hair drops behind her neck and then curves over the right shoulder. It kept looking very

● **Top Right** | The princess's all-seeing eyes give this beautiful build an ominous tone.

● **Bottom Right** | The hammerhead shark that lurks behind the statue provides some peril to the ocean scene.

"There's that saying that when you finish ninety percent of a project, you have ninety percent to go. That's exactly how I feel with every project that we work on."

weird because when you try and make something with gravity in a videogame that doesn't have any gravity, it becomes exponentially more difficult. Since we're so used to living in a world with gravity, there's one specific way everything needs to look. If you don't do it that way, everyone can recognize pretty quickly that it looks weird." Only in Minecraft could you find less frustration in creating a shark than in giving a haircut.

But Varuna managed to pull it off, and the princess is actually the part of the build that Thomas is most proud of. "I think we really nailed that!" If there's a secret to his team's success, it's to keep going and push yourself through the frustrations. "You need to develop some level of patience. It only looks good if you are able to sit down with it over several weeks, perfecting it little by little. If you expect it to be perfect after you've built something from scratch after a few days, you're going to get frustrated with the work that you're creating. You're just going to give up pretty easily, because it doesn't look exactly the way that you want it."

"Developing that patience is extremely crucial. We had four people working on this project, and it took them several weeks. The shading and the texturing – those are details that are really slow to add. There's that saying that when you finish ninety percent of a project, you have ninety percent to go. That's exactly how I feel with every project that we work on." So be a little patient with Varuna, and they'll reward you with a craft fit for a princess.

Created by Chloriz as part of his application to work at the studio, Agnoia Frieden depicts a majestic statue of a woman. She stands within a valley and is surrounded by derelict temples. Both the statue and the buildings are covered in vines, moss, and weeds, as if nature is taking back the sacred space. Chloriz got the job, little surprise when you consider how well it matches Varuna's skills as a studio in imitating organic subjects

BONUS BUILDS
Agnoia Frieden & Vigrid

Vigrid

The word Vigrid means "a place where battle surges" in Old Norse. The creators of this build, BlackKing and RenQeN, have used this inspiration to craft an island under attack. Not only is the village of Vigrid facing three pirate ships, it also has an angry T-rex and an erupting volcano to contend with. The crescent island is peppered with bright red buildings, which contrast against the greens, browns, and blues dramatically.

CHAPTER FOUR

Imperial City

By Rigolo and Comeon

The Imperial City by Rigolo and Comeon is a massive mish-mash of palaces, statues, buildings, and so much more. Inspired by several European cities, it should, by rights, be an incoherent mess. In fact, one of its architects might just argue that it is. But it's a tourist trap that truly has everything, a seemingly never-ending city so vast that backpacking through Europe might even be the faster option.

It was Rigolo's younger brother, Comeon, who introduced him to Minecraft, thinking it would appeal because of how much Rigolo enjoyed Lego. "There were never enough pieces, so the idea of being able to build something with unlimited material – I liked it." After a few practice projects to get acquainted with the game ("probably a horrible house or something," says Rigolo of his first build attempts), the brothers quickly decided to create a city. "We opened the map and we started with this project. The only build I ever made in Minecraft. I've never worked on anything else." Well, when the project covers this much ground, is there any need to?

There was no pre-production. Rigolo and his brother simply decided "to build something impressive without any plans. We started with a square, started to put buildings around it, and bit-by-bit we started building bigger and bigger, learning all the little tricks to be able to build things that are that big."

Plenty of Minecraft architects craft a building and call it a day. But Rigolo and Comeon, perhaps because they'd never made a plan, didn't know when to stop. Whenever one of them would build something, Rigolo would identify what it was missing, and construct that too.

● **Top** | Taking a lead from modern European cities, the brothers mixed architecture styles and periods to give credibility to their city.

● **Bottom Left** | Rigolo and Comeon added splashes of color to the buildings to invigorate the natural stone palette.

● **Bottom Right** | They even designed a sports arena fit for the Olympics!

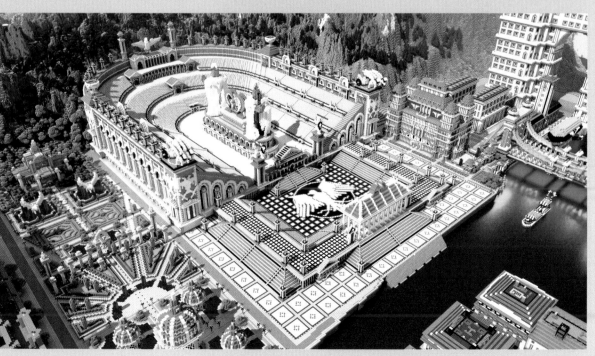

Fairly Inspired

World's fairs – celebrations of cutting-edge technology during the nineteenth and twentieth centuries – were a big influence on the Imperial City. But it's obviously too big a build to have just one influence; the brothers were inspired by architecture from all over the world. "There's a lot of Stalinist architecture. These very high, augmented buildings. You can see some of them in Moscow, where they mixed a lot of architectural elements from the baroque, neoclassical period." Other influences were a lot closer to home, such as the large domed French Pantheon. "We reproduced, a little bit, the environment we grew up in. I moved to Paris when I was eighteen and those big monuments are the kind of architecture you have around."

"There was an island with two rivers on either side, so that implied bridges." Bridges that needed to lead somewhere, so they built more streets. Streets that looked empty and unrealistic without buildings, so they built more of those. Buildings that needed neighboring buildings to connect to and ... well, you get the idea. Suddenly it's not hard to see where twelve to eighteen months disappeared to. When asked if there's any particular part of the world they drew architectural inspiration from, Rigolo answers, "Prague, Vienna, Paris, London,

● **Top** | The brothers designed many bridges to cross the rivers and canals, as well as boats.

● **Above Left** | They referenced several different architectual styles in the decor of the city, including Slavic and Venetian domes, plus Gothic facades.

● **Above Right** | Rigolo's efforts to make realistic streets are evident in the details, such as the street lamps, decorative trees and shrubs.

Brussels, Spain, Italy." Perhaps it would have been easier to ask which cities didn't inspire Rigolo. That all-of-Europe-and-the-kitchen-sink approach is evident in the variety the city boasts. You can enjoy clock towers worthy of Elizabeth Tower (commonly known as Big Ben) on one street, lush green gardens on another, and a fleet of sinister-looking tanks on the next. With so many varied influences and so much going on, how did they keep the city looking coherent?

"When you start looking at it, it's not that coherent!" admits Rigolo, explaining that the skeletons he built on the cliffs were an attempt to partially disguise that. "It was intended to be the remains of some ancient civilization that would have existed in this location before the city was even built. This was a period when people built all kinds of buildings with different styles. Especially as Europeans, we are used to seeing this kind of mix-up in the fabric of our cities. That's what

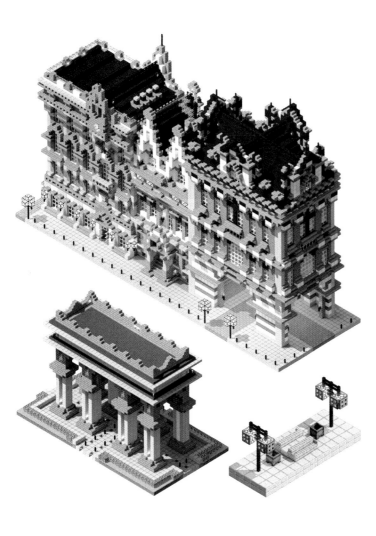

defines a city. You don't find many cities with completely uniform architecture. It's always layers of periods, history, taste, that have been accumulating."

It's a good argument for throwing in any building you can think of, but this process naturally lacks a clear finish line. How did they stay motivated? "You want to prove your worth and do better than others," says the wonderfully honest Rigolo. "You like what you've been doing so you feel motivated to expand it even more." While they were still working on it, they also shared the city on Planet Minecraft – one of the most popular community websites. The city was a huge hit, with the positive feedback leaving Rigolo and his brother "wanting to be even more successful." Cue another few dozen palaces ...

Rigolo and his brother built "about ninety percent" of the city as it is now, but when they opened it up to other players, they soon had other Minecraft architects volunteering to add builds of their own. "They showed us pictures of things they had done and we thought, 'Well, maybe.'" You'll see a few of those builds, such as a Gothic cathedral, in the final city. "But there's a lot of things that were built by others that we deleted because we didn't find them satisfactory."

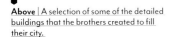

Above | A selection of some of the detailed buildings that the brothers created to fill their city.

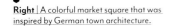

Right | A colorful market square that was inspired by German town architecture.

Sightseeing

Adding to their list of historical European architectural references – French, Italian, Russian, Spanish – Rigolo and Comeon were also inspired by British designs. "We took a lot of things from Victorian architecture. Factories that are very much inspired by nineteenth-century factories you can find in big English cities."

This Spread | As well as intricate building designs, Rigolo and Comeon also focused on other elements that make up a city, such as the vegetation, vehicles, harbors, landscapes, and the view from within the city.

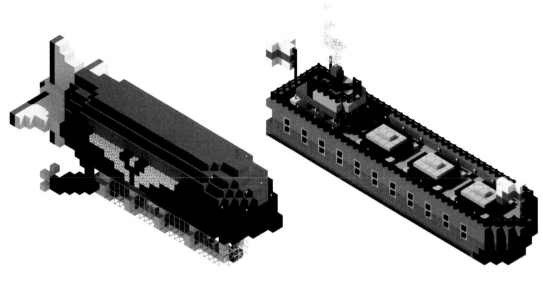

That may seem a little harsh, but these were the increasingly high standards Rigolo set for all of the city, especially as his own building skills improved. "There's a lot of buildings we came back to, to make them much more ornate than they were at the beginning. When you get started with building, you tend to make your facades very flat. There's not a lot of detail. We learned that you could add a lot of elements that would make it look more beautiful, more realistic, and more lively. So there are several buildings that were almost entirely redecorated." Those redesigned buildings were the lucky ones, as at least they got to stay. "I started several buildings that I didn't like. I erased them entirely."

Rigolo found it frustrating to spend tens of hours working on something only to throw it away (though he's quick to remind himself "it's just a game"). But having such a high bar has left him with a huge Minecraft city, acclaimed by the community. Surely he's happy with that? Well, when asked if he would change anything if he started the city today, he says "I would probably eliminate half of the buildings.

I don't find them up to par with the rest in terms of detailing, balance, and aesthetics. I would probably put in more houses and less big monuments. It's more motivating to build a palace than a house when you want to do something impressive. But that's actually something completely unrealistic in the city, that there are very few houses, streets, or shops." Whereas most would look upon this city and be blown away by its epic scope, Rigolo is more concerned with where you're supposed to buy your milk.

How did Rigolo decide when it was time to declare the city finished and stop building? "I couldn't tell you! We didn't decide to stop. And, actually, it remained unfinished. At some point we moved on to other things. I guess everyone likes to have a little pet project like this." After about eighteen months, Rigolo stopped playing Minecraft and deemed the city unfinished. But no city is ever really finished, even in the real world, and what he leaves behind is a testament to the scale and skill that two brothers, tired of running out of Lego bricks, can achieve.

CHAPTER FIVE

Earthsky

By natsu3012

Anyone who's had the dubious pleasure of living in a densely populated city will see Earthsky as a home away from home, because it so successfully captures the chaos of an urban jungle. Yet this chaos suits the tastes of the builder perfectly. Ambroise Guenole, a French architecture student who crafts under the name natsu3012, can't get enough of packing in more details, structures, blocks, ambitions, hopes, dreams, and probably another fifty things since you started reading this.

In their studies, natsu3012 likes to predict what cities will look like in the future. "For a utopian vision that would be beneficial for all, we imagine a natural city with a densified public transportation network, tall buildings, and rooftops that serve as public spaces. In my creative and aesthetic imagination, I would like the cities of tomorrow to look like my work, even if it would be unlivable in reality. It's a rather selfish dream after all. Depending on the path humanity takes, two scenarios are surely possible: a city of technology or a gigantic slum."

This builder's passion is so infectious that they might actually be able to sell you on the apocalypse. "I really like the atmosphere of an apocalyptic future. And Minecraft lends itself quite well to the realization of this kind of universe." However, for this vision of the future, natsu3012 planned something different. "I wanted to create a heap of buildings on top of a kind of floating base." That explains the "sky" part of the build title. It's a logical leap off the ground – when there's no more room on Earth for buildings, people are going to inevitably start looking skyward for somewhere to put their real estate.

Natsu3012 finds that their architecture studies help when planning such a lofty design. "I know how to conceive a project, develop it, and

"Depending on the path humanity takes, two scenarios are surely possible: a city of technology or a gigantic slum."

Sky Scraping

Natsu3012 spent about a month and a half constructing Earthsky, crediting the Minecraft: Java Edition plugin WorldEdit with helping save them a lot of time. As for the name of the city, they "had to display something on the main hologram" that gives the city a cold, blue slice of futuristic detail. "The name resulted from the intentions of the project. A city in the sky. A new land where people pile up and buildings rise to the sky." People pile up? Maybe best just to focus on the "buildings rise to the sky" part!

imagine the spaces at different scales." Definitely a helpful skill to have when a significant part of your structure is sky-bound. They also have access to "a repertoire of [architectural] knowledge that I have fun translating into the 'Minecraft language.'"

They're happy to give examples of architectural language when describing the beginnings of the city-building process. "I start by defining a 'sample' neighborhood, in which I develop my structural principles. Then I work on a more global scale with neutral volumes to compose the whole project." Neutral volumes, for those of us who

don't speak architect, are "simple volumes like cubes. It is a basic form that I come to rework afterward. I come to rework each zone in detail, so they are in coherence with the sample district." It's a smart approach – crafting a small, detailed portion so you always have something to go back to when reminding yourself what kind of city you're aesthetically aiming for.

Another smart method is natsu3012's creative use of color. "I don't use the blocks in the way Mojang intended. I use them for their color or texture. For example, I don't use dirt to make ground, I use it to

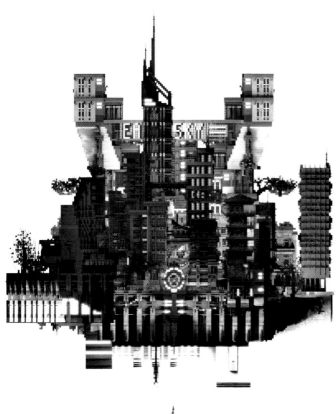

"To build on a large scale. I would advise first of all to have a solid foundation with very simple constructions. Understand that simplicity is not a defect, it allows a better reading."

make walls. I used a lot of different blocks. TNT, lapis lazuli ore … a block that I particularly like is mycelium, which I use very often." This process makes a pleasingly colorful city. "The goal is to break down the elements of a building. To bring out its materiality, its structure, to play with its gradations, and to simulate the effects of time – for example, paint that's steadily deteriorated – to reinforce the effects of light. Lighten the areas that are supposed to be light and darken the areas that are supposed to be darker." That may seem like an incredibly granular approach to each and every building, but it clearly delivers a great-looking build.

Natsu3012 also uses a popular building trick during the early stages of construction. They craft large structures out of colored wool blocks, then later replace them with the blocks they actually want. A smart way of getting "a better view on what you are going to do," as using placeholder blocks means you can quickly put together shapes for buildings or other structures. Which begs the question: Would you prefer to live in a house with walls made of wool or dirt?

When it comes to building designs, natsu3012 is interested in a wide variety of styles. "I enjoy making small constructions without great ambition. Small houses or other [projects] with multiple cultural inspirations while trying to improve my technique. Creating on a small scale is very rewarding. But Earthsky is quite representative of what I like to do most in this game – big, futuristic, and rather dystopian cities. It's an atmosphere I like."

This colossal city isn't even a third as ambitious as natsu3012 would like it to be. "Today if I was doing a similar project, I think the number of buildings would be doubled or even tripled, with paths everywhere, to really give the impression of a pileup. My texturing skills have improved so I think it would have big improvements on that side." Somebody change natsu3012's Minecraft password before they try to quadruple this city in size!

● **Left** | Without the light from the sun, Earthsky is lit from within. natsu3012 added different colored lights to enhance the busy city vibe.

● **Right** | The front-on perspective provides additional detail of the paths in the top left.

Below | natsu3012 designed a giant digital billboard to sit atop their city.

Bottom (Left to Right) | Rooftop gardens are a popular feature in modern builds that help to break up this metropolis; natsu3012 added colorful walls to their futuristic buildings.

Considering natsu3012's strong passion for large builds, it's somewhat surprising that they're actually a big advocate for the merits of keeping some things simple. "To build on a large scale, I would advise first of all to have a solid foundation with very simple constructions. Understand that simplicity is not a defect, it allows a better reading." It's how they've managed the tricky balance of making a city look deliberately claustrophobic without being an eyesore.

This meticulous builder likes to stress the importance of planning ahead. "You must have a guideline, an intention for what you plan to do will allow for a coherent project." And, of course, they had a

strong idea for their futuristic city ahead of beginning the building. "The plan for Earthsky was to make a 'finished' city on a limited surface, with buildings that are superimposed, but practical for the players. There's a multitude of paths at different levels that it is possible to take."

Although it's a floating city, natsu3012's Earthsky is surprisingly grounded in realism. "I wanted to design a walkable space, sometimes looking like alleys that go up and down, sometimes more like public places." This architect may have two different predictions for the future of humanity, but if they're the one planning it, we may be in safe hands.

CHAPTER SIX

Gothencrow
By Rajkkor

TEXTURE PACK
■ **Dokucraft Light** by Dokucraft

othic build Gothencrow is a proudly gloomy city of dark lore, featuring imposing, immensely detailed architecture that actually started as something significantly merrier. Gothencrow's architect Vincent Grenfeldt (known in the Minecraft community as Rajkkor) had just joined the Minecraft server and YouTube channel FyreUK. Three months into joining, his first project was to participate in one of their videos, but they wanted to populate it with more cities. "That's when I thought it would be nice to have a Gothic world in there. FyreUK hadn't done a Gothic-themed build yet." He then found out it was going to be a time-lapse video, something he hadn't worked on before.

Minecraft time-lapse videos show an entire build being constructed block-by-block, super sped up. They're excellent for seeing how teams put their builds together and incredibly addictive to watch. They also require a lot of answers before even placing a block. "What are the houses supposed to look like? What are the big builds in it? What is the style? What types of details are in there?"

Vincent soon discovered there would be limitations. "It turned out that apparently the details that I'd chosen for the build were a bit too complex for what the average person can do under a time-lapse. Because when you do a time-lapse, you have to take into consideration that not everybody has built for more than a year or so. It has to be a simple style, but look good enough. The purpose is to construct the map from the ground up." He eventually decided to just make Gothencrow a big world instead of a time-lapse.

Vincent took inspiration from the games he enjoyed playing. "At the time I was interested in *Dark Souls*. I really liked the old, European-inspired architecture in that game, so I thought it would be interesting to do something on that theme." For those who have never had the pleasure of dying over and over again, *Dark Souls* is a huge fantasy RPG by Japanese studio FromSoftware. Minecraft is slightly friendlier though, whereas *Dark Souls* is the weeping king of grimdark video game fantasy. It's full of imposing architecture that looks like you could cut yourself on it just from looking at it. Both clear inspirations for

Above | Although the title of the build is Gothic, the choice of blocks is surprisingly colorful. Vincent used prismarine to add a little vibrancy to his fantasy town.

"My friend introduced me to Minecraft. It was a long time ago!
I was around twelve. I started playing Survival mode and it was
some time later that I got into map-making."

● **Far Left** | The colors of the church stand out against the pale backdrop of the snowy mountains.

● **Above** | Vincent has perfected hiding majestic castles behind giant, wooded mountains.

● **Left** | What secrets might be discovered in this mysterious waterfall clearing?

Gothencrow, but Vincent was also influenced by "Gilneas, a city in *World of Warcraft*." Blizzard's long-running MMORPG is pretty dark in places too, but has a far more colorful and welcoming approach to fantasy. "The ambience and the environment came more from *World of Warcraft*, with the fairy-tale influences, and the architecture of choice became more like *Bloodborne* and *Dark Souls 3*. The entire theme of the city evolved as my interests changed."

This is somewhat inevitable when builders take on projects as large and time-consuming as this one. Naturally, as your build expands and time passes, your tastes and interests are going to develop and change too.

"People mature, but their tastes don't actually change that much. They want something similar to what they had as a kid, but more adapted for their current age." You can see the change in his interests in the loose narrative that Vincent constructed for Gothencrow too. There are pulpier ideas, like werewolf-summoning priests, but also more mature details, like the evolution of the city from a small village, and how that growth affected the building styles.

What inspired the name "Gothencrow"? The "Goth" part is pretty obvious. Vincent happily admits he had no knowledge of Gothic architecture when he pitched the build to FyreUK. "I just wanted to make something like Gilneas and *Dark Souls* and they said 'OK, so you're going for Gothic architecture?'" As for the "crow," you'll find insignias of these inky-feathered friends all over Vincent's city. "I think that was inspired by *Game of Thrones*," he says, pointing out they're a rightful cliché of the genre and provide a wonderfully spooky atmosphere. "If you ever have a haunted mansion, you'll always hear crow sounds in the background." But the third influence on the name was actually Gothenburg, the Swedish city, with its harbors and architecture inspiring a lot of Gothencrow's layout.

If he was building the city today, Vincent says he'd approach the design differently. "Gothencrow actually had a pretty planned layout. There is no city that is built like that naturally, at least not in Europe. Most cities start out as a village. OK, now we need a church. OK, now we need a market for the traders. Oh, it looks like we've got a new religion, going to have to make a corner for that. Oh, looks like the scientific revolution just happened, we're going to have to extend the university." His favorite part of the build is the king's castle. "That was the last build I did." Vincent sees the castle as a showcase of just how much his building skills had improved between joining FyreUK and starting Gothencrow. The more Vincent thought like an actual city planner, the more his build logically snapped together. "When I designed the castle, I tried to think about what parts would have been built first. What parts would have come later as the kingdom prospered? This is especially true for castles in general, where it will start as a fort, and as the village around the castle grows and walls are built, the need for the castle to be protected is diminished. I find that part of building history exciting."

As the main inspirations for Gothencrow were video games, it's not surprising to hear Vincent has been gaming since he was young. "My friend introduced me to Minecraft. It was a long time ago! I was around twelve. I started playing Survival mode and it was some time later that I got into map-making." Vincent might have stuck with Survival, had his

●
Below (Left to Right) | Everyday houses sit among this regal build; a giant eagle watches over the waterfall glade; a detailed cutout of the windmill and house.

●
Right | Vincent has created small villages within his Gothic build. Equipped with a small church and windmill, this cozy hamlet is overlooked by a beautiful tree in bloom.

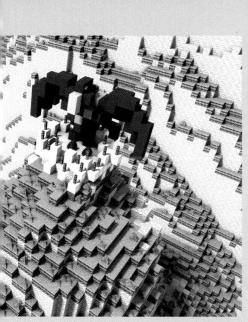

A Storied History

When designing a build, Vincent likes to think about the narrative, and the fortress with the small tower – to the right of the harbor – has a very detailed story behind it. "That is where the monster hunters have their headquarters and monster prison. I try to think about the history of the building, what part of it was built first, what parts of it were added later as a need for expansion for whatever purpose the owner of the building would have. In the case of the monster hunters, the tower is an extension of the lower building, which served as the guards' official prison until they discovered the monsters and needed to develop a special tower to hold the monsters in it."

school not suddenly given him a great incentive to try a new approach. "There was this competition where we were tasked with creating an adventure map." The only rule of the competition was that the map had to have an underwater theme. "We made an adventure map about a character that's very similar to Captain Nemo from *Twenty Thousand Leagues Under the Sea* by Jules Verne. The storyline was about a guy finding an underwater city. So it wasn't very original! But it was more about the actual building, of actually constructing the world around it. We built our interpretation of Atlantis."

Vincent and his team's elusive underwater city won the competition. But all Vincent can see now is his old build's flaws. "Minecraft at the time didn't have water that wraps around the blocks," he explains, of a build constructed long before The Update Aquatic improved Minecraft's oceans. "If you went down and put slabs underwater, it would create a vacuum or air pocket around the parts of the slab that weren't covered by the full block. So it didn't look that good. Then we got complaints from one guy who built an underwater base who said his build was cooler and therefore he should've won. I was like 'OK, well, I better improve.'"

That sore loser was a surprisingly good motivator for Vincent. He started looking for a Minecraft server to join that would help improve his building skills, as he'd definitely caught the map-making bug. "I really liked to make *Tomb Raider-* or Indiana Jones-style maps, where you go to these epic locations and then you explore them and discover old architecture there."

That's when Vincent discovered his ideal server. "I liked FyreUK a lot so thought it'd be nice to join them, do some building, and find a more concentrated building community, where one could talk with other builders and learn from them." As well as Gothencrow, Vincent worked on other FyreUK projects, such as Christmas Timelapse 2013, Elven Port 2.0, plus several time-lapse projects, such as Kirtipur. FyreUK have sadly since closed their digital doors, but he clearly learned a lot, constructing a Gothic delight that's as sinister as it is spectacular.

Above | This picturesque town is reminiscent of Bavarian and Eastern European cities, packed with colorful tiles and wooden detailing.

"Gothencrow actually had a pretty planned layout. There is no city that is built like that naturally, at least not in Europe. Most cities start out as a village."

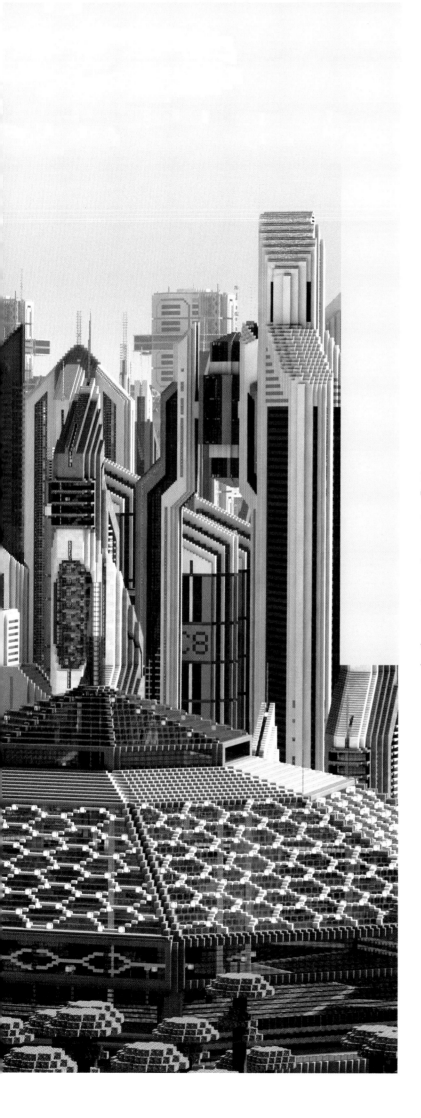

CHAPTER SEVEN

Future City
By Zeemo

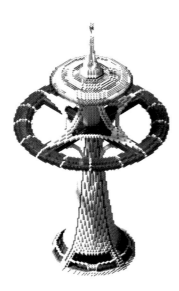

It was early days for the Minecraft community when French builder Zeemo started playing, but he was still impressed by what he saw. "When I discovered the constructions of the other builders, I was fascinated by the beauty of the castles and monuments." But most of the community builds he saw were medieval-style. "The few existing cities did not have anything futuristic in my eyes." So Zeemo decided he'd be the one to fill that gap. "I looked at several futuristic-looking games and images and gave it a go."

It's a clever approach. Often identifying what's missing from the builds you're seeing in the community and filling that gap can be a great way to get more attention on your project – after all, there will be nothing else like it. Zeemo cites games like *Halo* and *Mass Effect* as influences, along with the futuristic movie franchises *Star Trek* and *Star Wars*. "I looked at several futuristic images for inspiration. Then I looked for how I could recreate these buildings in Minecraft, figured out what blocks to choose and how to do it. Over time, I ended up developing my own style."

Of course, the artists who worked on *Star Wars* and *Mass Effect* had access to some of the most cutting-edge technology for making digital landscapes. Whereas Zeemo started Future City around 2014, and was using the Xbox 360 version of Minecraft, significantly more limited than the versions available today (the lack of colored glass blocks was particularly frustrating when building a city of skyscrapers). "The blocks in vanilla Minecraft [a version that hasn't been modded] were a good match for a medieval world, but not really a futuristic world," which is likely why so many early community builds were so focused on the medieval era. If Zeemo wanted a new-age city, he was going to have to get creative.

"There's realism despite the coarseness of the blocks. The freedom offered by the game and the calm it brings, all accompanied by music that is both motivating and relaxing."

This Spread | As well as building an incredibly detailed city, Zeemo spent time creating a selection of aircraft to populate his sky.

"I looked at several futuristic images for inspiration. Then I looked for how I could recreate these buildings in Minecraft, figured out what blocks to choose and how to do it. Over time, I ended up developing my own style."

Luckily, that was far from a problem. "I started by using the clay and snow blocks, arranged with blocks of wool of all colors. To vary the shades of gray, I even had to use the cobblestone block." Which means this city of tomorrow's foundations are some of Minecraft's earliest, most primitive cubes.

"When the quartz block arrived … it was my block par excellence. I finally had access to the quartz stairs." And he made great use of it. As for the lack of colored glass on Xbox 360, Zeemo found a workaround by using ice blocks as windows, making use of their blue shade. It's proof that sometimes restrictions can be good for creativity.

When you've got unlimited access to resources, it can be tempting to try to use everything, leading to a design that looks cluttered or overly busy. By having such a small color palette, Zeemo's Future City looks strikingly consistent throughout as well as thoughtfully realistic.

Zeemo cites the Burj Khalifa, a skyscraper in Dubai, as an influence. "I love the futuristic and uncluttered design of many of the buildings I've made. I often finished my buildings with an angular roof … an often-luminous mast or some kind of antenna. Like the Burj Khalifa, the top floors are there to complete the design, to increase the presence of the building."

Under the Dome

Zeemo's Future City looks inviting, but it's far from a utopia. You'll find a base with several tanks, and even more sophisticated defenses. "The most elaborate and also the most striking section is Area 2-57, a gigantic glass dome." Area 2-57 is an experimental, militarized base, created as a precaution in the event of attack from the city's enemies. "This dome contains several very futuristic buildings, but also four cylindrical modules with lasers." It's one of the most complex areas of the city, and the part that Zeemo is most proud of.

Although Zeemo enjoys modern architecture, he's not a fan of more recent design trends in building design. "I am not from the school of building entirely in glass panels. I love an illuminated concrete column decked out in glass. Architects today lack originality. Their achievements are often anonymous, without novelties, practically all identical, and boring to look at. I hate the organic style of buildings and the curves that lead nowhere."

Zeemo was introduced to the game by his eleven-year-old son, who was playing the beta version of Minecraft around 2011. "I found the game

to be really fun, but there was no Creative mode at the time." Luckily, he wouldn't have to wait for some long-distant future civilization to invent one, as Creative mode was one of the earliest Minecraft additions. As soon as his son announced that Creative mode had been added, Zeemo dived back in.

What was it he liked about the game? "The first thing that struck me … were the blocks that exploded when they're mined." Ah, he means literally struck him. No wonder he prefers Creative mode. Traitorous blocks aside, Zeemo was also taken by "the immensity of the map in all

● **Above** | In such a metropolitan build, you would expect the buildings to dominate, but Zeemo has made sure to balance the cold steel and glass with plenty of green areas.

● **Right** | Zeemo has filled his futuristic cityscape with iconic structures, reminiscent of 1960s spacefaring designs.

Oh So Arty

Art Deco, the style that inspired Zeemo's city, originated in Paris and spread all the way to New York. "Art Deco" is short for *Arts Décoratifs*, which comes from *Exposition Internationale des Arts Décoratifs et Industriels Modernes* – a French World's Fair held in 1925 that brought international attention to the style. Those 1920s New York skyscrapers had a huge influence on pop culture's idea of a futuristic city. A lot of movies, TV shows, and games use Art Deco motifs in their designs. One early example is Fritz Lang's 1927 movie *Metropolis*, which merged modern technology with opulence to create a futurist setting that still inspires Hollywood today.

"If I had to start over I would do it differently. I would avoid certain styles of buildings to keep only those that are the most representative of a futuristic city. The style would be much more homogenous and the development of the city better planned."

directions, with its biomes, mountains, valleys, lakes, rivers, rain, and snow. There's realism despite the coarseness of the blocks. The freedom offered by the game and the calm it brings, all accompanied by music that is both motivating and relaxing."

He's a big fan of experimentation and there's a lot of room for that in Minecraft. "Sometimes you try something, and the effect works and other times it doesn't. When you don't like it, don't be afraid to start over and try something else, a different style, a different color, or a different material." Keep trying new things until you find what works for you.

Eight years after starting, Zeemo continues to work on Future City. He's since converted the build from Xbox 360 to the PC version ("When I learned of the existence of shaders on a PC, I gradually abandoned the game console") and it's the PC version you're admiring here. His city is no longer an outlier amongst all the fantasy or medieval builds, as the Minecraft community has fully embraced science fiction. "I had no other futuristic city as reference," Zeemo remembers from when he started the project. "If I had to start over I would do it differently. I would avoid certain styles of buildings to keep only those that are the most representative of a futuristic city. The style would be much more homogenous and the development of the city better planned."

It's refreshingly optimistic to take a stroll through Zeemo's Future City. The clever use of a limited color palette to depict an efficient vision of tomorrow, vast swathes of thriving nature, and easy access to both subways and teleporters throughout the city. True, the presence of teleporters should make that subway completely obsolete, but let's leave such petty pedantics in the past.

CHAPTER EIGHT

Kythor Landscape

By qwryzu

TEXTURE PACK
- **Faithful 1.8** by Faithful
- **Better Terracotta Colors for Faithful** by qwryzu

Minecraft is a proudly surreal collection of colorful cubes, each one an unabashed love letter to retro 2D pixel art. There's no attempt at photorealism here. But in Kythor Landscape, builder qwryzu has crafted terrains that actually resemble the real-world landscapes that inspired them. Now a game that deliberately rejected realism looks realistic.

Minecraft is far from qwryzu's only passion. "Terrain [in my builds] is my main focus because I'm from the American Southwest and nature has always been a huge part of my life. I love traveling, hiking, camping, and just generally exploring the huge diversity of biomes that my home region has to offer. If I had all the money in the world, I would do nothing but explore."

Unfortunately, qwryzu doesn't have all the money in the world. Since he couldn't "spend all my time in the outdoors" he started finding clever virtual ways to enjoy the outside while staying indoors. "I began Minecraft terraforming as a way to enjoy pretty landscapes without actually having to go in real life. I would build cozy cabins in the woods but was never happy with how the default mountains looked, so I began trying to make my own."

A cabin is a great choice of cozy craft. After all, who doesn't want to put their feet up by a roaring fire, sit back in a comfy chair (or, by Minecraft rules, stand next to a comfy chair), and relax? Well, qwryzu, apparently, who found himself focusing on the landscapes surrounding his constructions rather than the builds themselves. "Every time I made a new map, I would build on it for an hour, but could do nothing except see every way I wanted to improve the terrain with my next one. Eventually I stopped even building on my maps because I just wanted to keep on making the landscapes."

Luckily, he'd found a game with unlimited options. The bigger a player's imagination, the bigger the build. "Minecraft is a game that markets itself with endless possibilities, and might be the only game I've ever found that truly means it. I love the freedom to be able to build whatever I want. I'm sitting here making pretty mountains while there's other people out there making actual computers in the game. Minecraft's community has also created one of the most vibrant modding/data pack scenes in the entire gaming world, which means that there's never really any shortage of things to do even when you're out of ideas for vanilla Minecraft."

Top | The serpentine river winds its way through the luscious jungle valley.

"I love traveling, hiking, camping, and just generally exploring the huge diversity of biomes that my home region has to offer."

Southern Lows

"The southwest continent was inspired by New Zealand and the Pacific Northwest of the US, both of which have awesome, snow-capped volcanoes surrounded by beautiful forests. The southeast continent (pictured here) was vaguely inspired by a mix of southern Utah and the high steppes of central Asia, though I deviated a lot from those inspirations in the final build. The toughest part to build was the southeast continent in the transition zone between the red deserts and the grassy plains. I probably went through twenty different iterations of block choice because I couldn't make anything look natural, no matter how hard I tried. Eventually, I settled on the current look and I'm happy with the final product. But it was certainly frustrating to get there!"

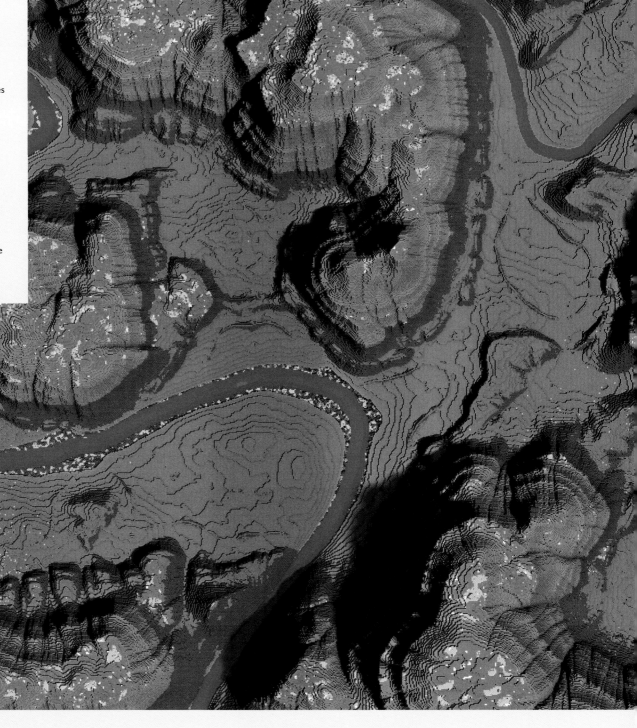

Qwryzu is a part of that scene, using mods to make the best landscapes possible. But first he spends some initial hours researching "Google Earth and landscape photography accounts on Instagram to find locations that strike inspiration," and also snapping some reference photos of his own. Then he draws a sketch of the general layout of the world once he has an idea of what he wants to include in his build. Finally, it's time to break out the building tools.

"The shape of my terrain is done almost entirely in software called WorldMachine," he explains. WorldMachine is very popular with the terraforming community and it's easy to see why. It lets you craft

"I love the freedom to be able to build whatever I want. I'm sitting here making pretty mountains while there's other people out there making actual computers in the game."

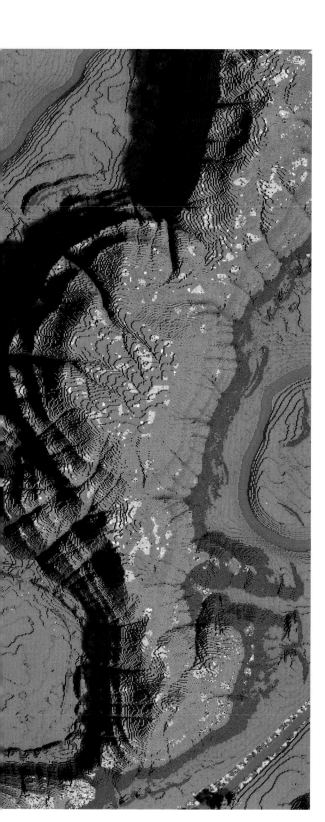

● **This Spread** | An aerial view captures the scale and majesty of this vibrant landscape.

realistic CG landscapes, using algorithms to simulate how terrain should behave in real life. Its erosion options alone are, appropriately enough, wonderfully deep. Would you rather simulate flow-based erosion, thermal erosion, or bury your terrain in snow, perfect for a nice cozy cabin? WorldMachine's initial difficulty curve can feel as jagged as an actual eroded cliff, but it's a great tool for those wanting to take their Minecraft terraforming skills up a few peaks. Outside of Minecraft, qwryzu is an aspiring geologist, one who uses the software to simulate processes that he learns about in his classes.

WorldMachine is also only the start of qwryzu's crafting process. He's also a big fan of WorldPainter. "WorldPainter is free software for creating Minecraft landscapes that is, without a doubt, the most important external tool in the building community, not just amongst terraformers. I have to do a lot of work in WorldPainter too, to apply layers like trees, plants, rivers, and some block selections that I didn't do with WorldMachine."

WorldPainter isn't quite as complex as WorldMachine. It lets you freely paint large sections of terrain, much like many digital painting or photo-editing tools. Vanilla Minecraft locks players into a first-person perspective, so a lot of players use WorldPainter or WorldMachine to get a wide overview of their builds. No wonder qwryzu holds them both in such high esteem.

His tools selected, now comes the hard part: putting all that terrain together. This "arduous process," as qwryzu describes it, involves a lot of jumping between WorldMachine and WorldPainter, making constant changes, and "exporting small chunks of the map at a time to see how they look." It's far from a fast approach, especially as qwryzu's also adamant that his landscapes are consistent with Minecraft's style. "I try to keep my block choices relatively simple and Survival-friendly. It's mostly a mix of stone and terracotta, with grass too, where appropriate. Stone and darker terracottas or concrete like cyan, brown, and black are great for big, tall mountains like the Himalaya or Rockies, but other colors like white, yellow, and brown terracotta can make awesome desert mountains, like the Atlas Mountains. My custom texture pack [featured here] also adjusts terracotta colors to more closely represent the colors of sedimentary rocks found in the Colorado Plateau region of the United States. A lot of terraformers use blocks of a certain color that look awesome in renders, such as coral and sponges, but they don't make any sense for Survival."

Qwryzu's love of Minecraft was born from playing the game in Survival: he was introduced to an early version of Minecraft in 2010. "I recall only being able to mine stone blocks and place torches and that alone was enough to catch my interest." Over a decade is a long time to spend with a game, but he had a personal directive that provided limitless challenges. "I want all my maps to make sense for in-game use as well, not just renders." Being able to create a realistic build, while still keeping it recognizably Minecraft, is an admirable achievement. Advanced as his terrain builds have become, it's nice to see nothing's eroded qwryzu's enthusiasm for the Survival game he fell for all those years ago.

"I recall only being able to mine stone blocks and place torches and that alone was enough to catch my interest."

● **Left** | To create the impression of a gentle hillside, qwryzu stacked smaller layers of blocks on top of each other.

● **Below** | This method is evident on a much grander scale in the builder's ambitious mountainscapes.

Northern Peaks

"The inspiration for this map came from looking at the Atlas Mountains in the northwest Sahara Desert of Africa. Being from the American Southwest, tall, snow-capped peaks above deserts have always been incredible to me. But the Atlas Mountains sitting above the endless dunes of the Sahara particularly captured my imagination."

Those northwest desert mountains and dunes are the part of the build qwryzu's the most proud of. "The oasis at the base of the mountains where you can see dunes, palm trees, and snow-capped peaks is without a doubt my favorite area on the whole map. It's a rare thing that I have an idea in my head and end up with a product that looks exactly how I dreamed!"

Tips & Tricks

01
Make a Plan

Planning is the most important lesson Zeemo learned while building his Future City, and his main advice to builders. "Before starting to build a building or a monument, you have to have in mind a certain vision of how you are going to proceed; architectural style, scale, choice of blocks, colors, lighting. On a map with a biome, the location of buildings is also important."

02
Install Helpful Mods

WorldEdit – the software that Maggie Gondo Wardoyo and Arkha Satya Taruna utilized when crafting their Moon Destruction build – is one of the best Minecraft mods. It lets you select and change several blocks at once, replace existing blocks without having to individually remove them and replace them one by one, and lots of other handy shortcuts that make building on such a large scale far easier. Some players don't like using mods but it's similar to how you'll only find a beautiful render of a build if the build itself looked good in the first place. Using tools like this can be a great way to save time and encourage you to experiment more.

03
Use the Right Tools

VoxelSniper is a Minecraft mod for world editing that Thomas Sulikowski, of Varuna, is a big fan of. "It's a great tool that allows us to push and pull certain blocks forward and backward. Probably one of the most useful tools when creating organics." Organics such as giant dinosaurs or human faces. It allowed them to move blocks around – "like how a sculptor creates a sculpture in the real world" – an enormous help for getting the trickier parts of a human face right. "Being able to push and pull blocks allows us to mold the shape of the jawline, for example."

04
Meet the Community

Arkha recommends reaching out to other players. "I think the Minecraft community is one of the most active and largest gaming communities in the world. We can share our maps, texture packs, skins, and Minecraft tips and tricks. I've met many people through Minecraft, including my best friend and partners on Minecraft projects."

05
Pick a Palette

"This is part of the one- or two-week planning process," Thomas Sulikowski explains. "We'll go into a Minecraft world and start placing blocks that we think are nice color combinations. For a smaller project these can be twenty to twenty-five blocks. For a bigger project it can run to over a hundred blocks. For Aqua Princess, we really wanted to add these pops of color, because the reference image we got was super colorful. The colors were very punchy, very in your face. We wanted to mimic that in Minecraft, so we wanted to make sure that we picked the right blocks."

06
Mods Are Your Friends

Conquest Reforged – used by Minecrafttalsi to make the Rich Medieval Town – is one of the most popular Minecraft mods and it's not hard to see why. Boasting over 12,000 blocks, extra textures, particle effects, and much more, it's particularly popular with historical and fantasy builders. Likely because the bleak medieval age – where rumor has it you couldn't even download Minecraft then – looks a lot more inviting through the rose-tinted glasses that Conquest Reforged's terrific toolset offers.

01.

02.

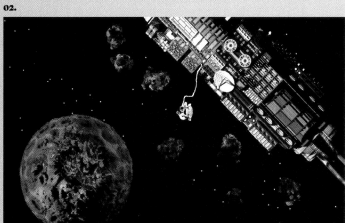

03.

 04 | Arkha and Maggie's Cyber City celebrates the future of humanity, a fitting scene from builders who believe in community.

● 05 | Varuna recommend choosing your blocks ahead of the build so you can find a palette that pops, like their vibrant blue tang.

● 06 | Minecrafttalsi finds Conquest Reforged provides excellent tools for his vast medieval world.

04.

05.

06.

● 01 | An aerial shot of Future City by Zeemo, who likes to plan out his builds in advance.

● 02 | To create their explosive moon, Maggie Gondo Wardoyo and Arkha Satya Taruna used WorldEdit.

● 03 | Varuna found VoxelSniper helped create realistic-looking organic matter, such as the beautiful statue's face in Agnoia Frieden.

CHAPTER NINE

The Sea Diary
By Junghan Kim

Short of visiting an aquarium with two kaleidoscopes glued to your eyes, you're unlikely to see a more colorful depiction of underwater life than The Sea Diary. A lush collection of aquatic creatures, mysterious temples, and almost every hue that Minecraft has to offer. It's enough to make you wonder why you even bother living on land in the first place. Surely breathable air is a small price to pay to live somewhere this pretty?

Remarkably, The Sea Diary wasn't even intended to be a sea build at all. Its architect, Junghan Kim, began this project desiring to create something simple. "I wanted to build something that could have existed in the past, like Aztec civilization." As a starting point, Aztec architecture lends itself nicely to Minecraft, with its chunky pyramids and grand scale. Junghan's design then developed into the vibrant neoclassical settlement that's seen here.

"How it came to be in the sea is that I thought there must have been topographical changes to the old civilization over time. I knew that the sea level was changing." Instead of researching those actual topographical changes, Junghan was suddenly inspired to take a more speculative approach. What would a civilization that had been completely consumed by the ocean look like?

Initially, Junghan wanted to build a city that was subjected to shallow waters, to give the impression that these elaborate temples were slowly sinking into the sea. An interesting premise, one that would certainly be challenging to depict in Minecraft. But Junghan had a change of heart and eventually decided to submerge the entire build. It wasn't the challenge that stopped him – the mysteries a deeper ocean offered simply enthralled him more. "I became interested in the deep sea because we never know what's happening down there. It's scary but interesting. So I wanted to make a work that explored this subject."

Top | This top-down view draws attention to the color palette used by Junghan Kim. An underwater rainbow!

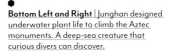

Bottom Left and Right | Junghan designed underwater plant life to climb the Aztec monuments. A deep-sea creature that curious divers can discover.

"I became interested in the deep sea because we never know what's happening down there. It's scary but interesting. So I wanted to make a work that explored this subject."

"Everything was produced mainly using creativity rather than research; I thought more about how to make it look better."

What isn't happening down there? Junghan seems to have had an allergy to dead space, cramming his build with as many sea creatures, temples, plants, and even bubbles (a lovely touch that enhances the underwater feel) as possible. "Scary but interesting" may not seem an immediately obvious design philosophy, given the luscious color palette that makes Junghan's ocean look so welcoming, but you don't have to explore for long to uncover more sinister details, like a gigantic shark with unfortunately-also-gigantic teeth. Swim inside its mouth (go for it, you'll probably be fine), and you'll discover several doomed fish, frozen in the moment before they become shark lunch. Junghan looked up real fish shapes, sizes, and ecosystems to help craft these creatures of the deep, but realism was never the priority. "Everything was produced mainly using creativity rather than research; I thought more about how to make it look better."

Real-world inspirations can be helpful, but Junghan was more influenced by other Minecraft builds. "I've seen sea works by other artists. The most impressive map was Deep Sea by Blockworks." Deep Sea depicts an underwater laboratory, with a significantly more muted, colder color palette than The Sea Diary. It certainly meets Junghan's "scary but interesting" criteria. Describing Deep Sea as a "really amazing arrangement," Junghan referred to it a lot when crafting his own map. He didn't try to recreate another build team's work, but admired it and used what they'd built as the foundation for his own original approach to the ocean. In this case, a complex structure that sprawls across the dark depths of the ocean, teeming with life to be explored.

Left | Who wants to live in an orange submarine?

Above | Pops of yellow and orange help the darker parts of the city stand out.

Junghan is quick to praise other builders, but it's his own building abilities that he hasn't always been so quick to praise. "I started this work when I really thought I lacked architectural skills. Many of my friends were really good at architecture and I really thought that I was lacking compared to them." It's easy to scoff at that claim now, when you see his finished build, but negatively comparing yourself to others is sadly a very effective way to demotivate yourself out of building anything at all. Thankfully, instead of giving up, Junghan made a promise to himself to stick with The Sea Diary.

That wasn't always an easy promise to keep. "It was really hard to spend so much time alone making everything in high detail and quality. I actually tried to change the story and the theme several times. They'd change based on my mood and feelings in the moment, forcing me to spend many hours editing the buildings and the terrain. To be honest, it was stressful to build sometimes."

The frustrations of endlessly adjusting a build of this scale, however, eventually gave Junghan the theme and story he'd stick to. "At some

● **Above** | Zebra fish are one of the many types of sea creature that Junghan has carefully recreated.

● **Top and Right** | An enormous shark swoops down on the unsuspecting school of fish.

Worth Sea-ing

When he was developing The Sea Diary, Junghan worked with two skilled artists, LNeoX and LordKingCrown. Their original renders of the build were essential in helping him find an audience. "A good render brings the build to life," LNeoX says. "It captures a certain perspective that should be seen by the viewer. With big builds it is impossible to catch the size and atmosphere in-game. This is where renders become really important."

"Just like with any book or movie, if the trailer or cover art does not look interesting, people will not care about it," explains LordKingCrown, on the importance of a good render. "Some builds look really good in default Minecraft, but with a render, you can really see all of the details. You can help create the vision the original builder had." A good render can't make a bad Minecraft build look pretty, but a great render can make a beautiful Minecraft build reach new heights! Or in this case, new depths.

"I'm proud that I wanted to do this and I kept my promise with myself by completing it. I completed this because I wanted to do it. That's the happiest thing."

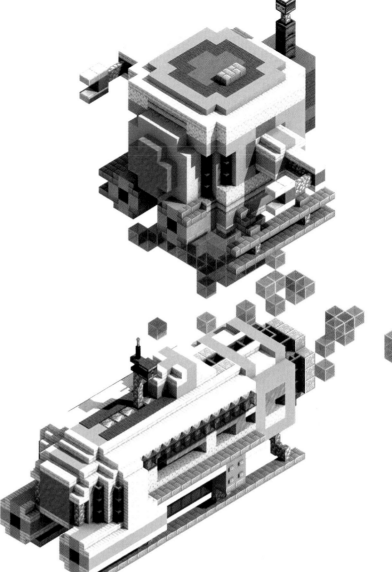

point I thought all of the moments I spent building this map could be the story itself, and this thought led me to create the 'sea diary.'" See, the titular diary isn't a book full of personal secrets that someone's unfortunately dropped in the ocean. The build actually is a diary. Junghan considers different parts of this underwater scene to be reflections of where he was emotionally when putting them together. It's an ambiguous narrative open to interpretation as you swim through it and explore. What was Junghan feeling when he crafted that wild-eyed shark?

When Junghan shared The Sea Diary online – a risky move, as it was technically his diary – the community immediately praised it for the attention to detail and fresh approach to what a Minecraft ocean could be. Delightfully, it's praise that seems to have sunk in. "Through this work, I think I recovered my self-esteem and confidence a lot. More importantly, I'm proud that I wanted to do this and I kept my promise with myself by completing it. I completed this because I wanted to do it. That's the happiest thing." It took over three years of crafting, but now Junghan is left with an outstanding build, renewed confidence, and a diary that's essential reading.

This Spread | As well as sea creatures, Junghan crafted underwater vehicles for his imaginary citizens to travel in.

Something Fishy

You'll find twenty-eight different types of fish in The Sea Diary, including a school of honey block-based creatures that are no doubt as delicious as they are adorable. Junghan used glass panes to create the impression of a tail that gradually fades out. A honey block with two buttons for eyes proves that sometimes the simplest ideas are the most effective. Junghan has made a glowstone block the "body" of this fish, ensuring the depths of his build are full of light sources, even without renders. His build embodies Minecraft as an interactive medium; it's a beautiful subject for screenshots but is also fun to explore. After a few minutes of swimming around their home, you'll envy these fantastic fish! Well, except maybe the ones in the shark's jaws.

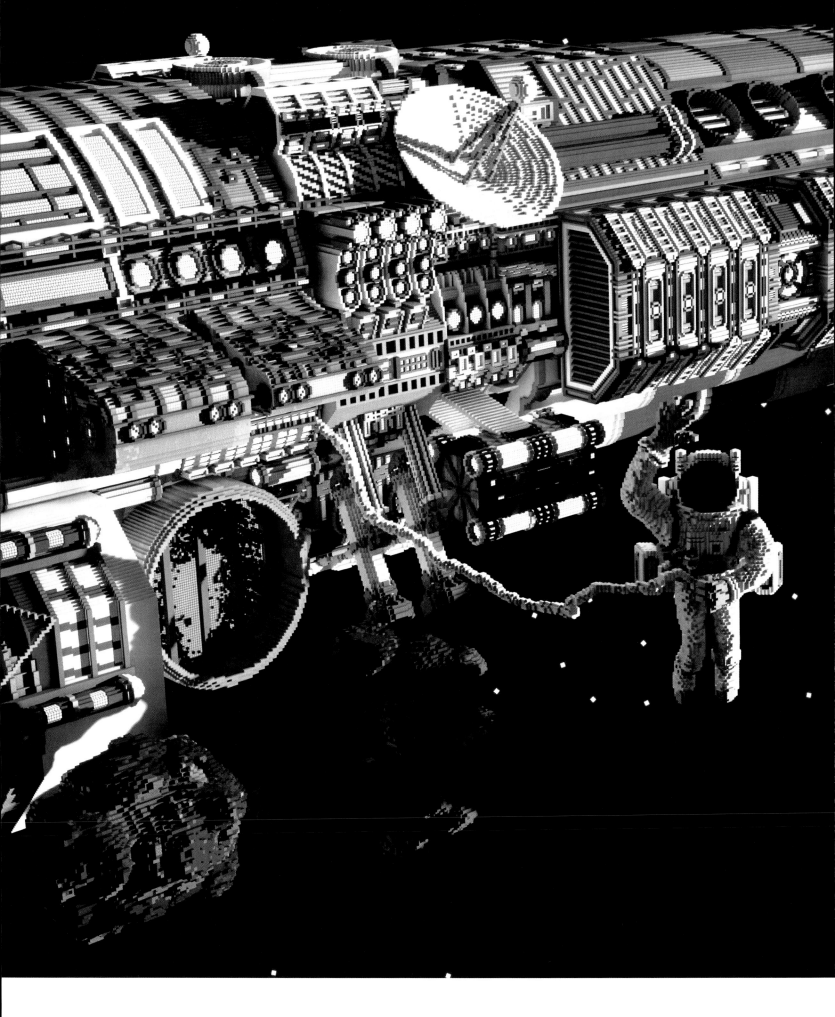

CHAPTER TEN

Moon Destruction, Front-Line in Space

By Maggie Gondo Wardoyo & Arkha Satya Taruna

Is there anything in Minecraft more gleefully disrespected by the building community than the height limit? When Moon Destruction was created, builders weren't supposed to exceed 256 blocks, but space-loving crafters couldn't care less about such earthbound limitations. They've happily shot for the stars. After all, what could possibly go wrong?

Well, quite a lot, according to the disaster depicted in Moon Destruction, Front-Line in Space. It's essentially three builds in one: the ship would look impressive in any space game, never mind a crafting one that is mostly based on firm ground; the lone astronaut, tenuously tethered to the spaceship, who deserves a commendation for staying cool in a crisis; and a not-so-friendly-looking meteor strike. The build depicts the kind of accident that actual astronauts probably lose sleep about, and according to co-builders Maggie Gondo Wardoyo and Arkha Satya Taruna, that part of the build actually was an accident.

"I copied a part of the moon's surface a bit too deep with WorldEdit," explains Maggie, on how that explosion came to be. "Arkha was better at making shapes. He had an idea to change the surface into a destroyed part with meteors hitting it."

Both hailing from Indonesia, Maggie and Arkha are a building duo that have gone on to have separate careers with Minecraft, even though Maggie wasn't completely sold on the building game at first. "To be honest, I thought Minecraft wasn't a good game because of the graphics – please don't feel offended – now I don't think so at all!" Even with that perception, it didn't stop her and Arkha from building something truly magnificent.

Looks aside, it was the game's potential and variety that sold her on it eventually. "You can literally do anything here, from building, programming, playing minigames on servers, or just enjoy surviving the nights. This is why I think Minecraft is more than just a game. It's also a medium to deliver messages to people through art."

"The game is like a canvas for me," says Arkha, also a big fan of Minecraft's artistic potential. "I can do anything and create anything that I can't do in the real world." True, but crafting an entire moon isn't exactly an easy thing to create in Minecraft either. "We made the moon entirely in-game and it took a lot of time because it's so big," says Maggie. The sheer scale meant they couldn't even get a

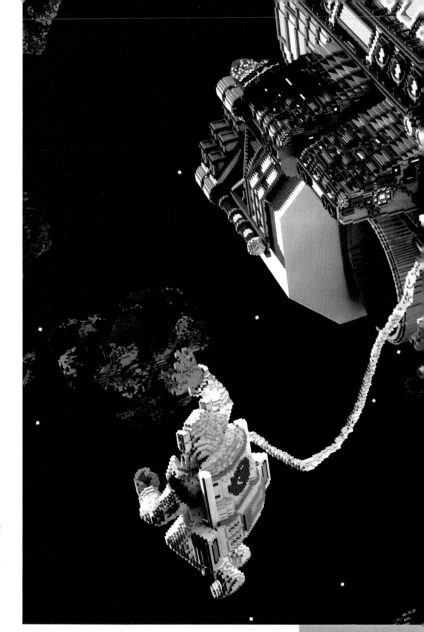

Left | Maggie and Arkha wanted their ship to be packed with realistic details, such as radar dishes, windows, and industrial-looking tech.

Bottom Left | An impressive explosion.

Bottom Right | The front perspective provides a stark contrast in color palettes, from the warm hues of the fire to the cold shades of the ship.

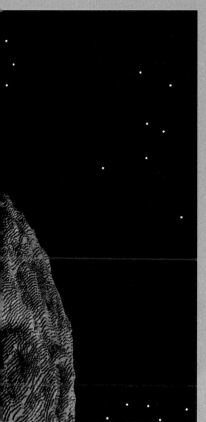

Star Stories

After Maggie Gondo Wardoyo's accident in WorldEdit left a crater in their moon, the duo needed to come up with a new scenario for their build. "We got the idea that the moon was hit by a meteorite and it affected life on Earth. Therefore, they sent some people to the moon to see and analyze what was really happening." Probably a good thing that you famously can't hear screaming in space, as they're in for one hell of a shock.

● **This Page** | A friendly astronaut, whose space suit was influenced by the movie *Interstellar.*

● **Right** | Maggie and Arkha utilized shape language in their flame designs to help distinguish between the controlled thrust of the ship and the chaotic fires of the moon explosion.

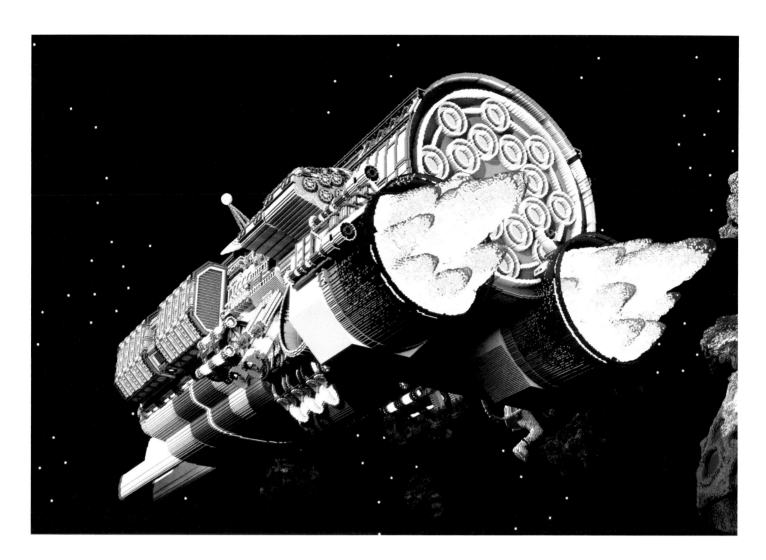

full view of the moon while they were building it. "We had to keep guessing when building if it looked decent enough!" she says. Despite those limitations, their final sphere of carefully constructed craters culminated in a moon that ranks a few notches higher than "decent enough." Still, it's nice to know that when these builders left for space, they kept their egos on the ground.

Destroying the actual moon might have been an easier project than putting this Moon Destruction build together. When asked what the hardest part to build was, the duo essentially said all of it. "Every part had different challenges and almost everything required the same level of attention," says Maggie. Special attention, however, went to the spaceship that dominates the scene. "We made it with extra effort. It took a really long time because of its size. Plus, we needed to make sure the details went well with the ship's shape and that the composition wouldn't be too crowded with the moon and astronaut there." So they had to juggle building a vast spaceship and keep in mind that it shouldn't take away from the rest of the scene. It's now all the more impressive that the astronaut doesn't have their head in their hands in despair.

One thing the endless black backdrop of space can hurt is your sense of scale. After all, the Death Star only looks threatening when you have a puny planet in the frame, which helps to showcase that deadly ship's terrifying might. Maggie and Arkha achieve this both with the astronaut outside the ship (also giving a nice human touch to a build

"It's a very agile method. It changed all the time ... and at the end I have something. Something that was absolutely out of control at the beginning!"

that could risk being too cold and clinical) and by going big on both the scale and the details. The spaceship is covered in unique patterns, tech, windows, a radar dish, and more, all in a coherent design. It's a star-bound vessel with a great sense of sheer bulk.

"We looked at scenes in the movie *Interstellar*," explains Maggie, of the research the duo did for their spaceship. "We looked for images of rockets and space stations to get a good grasp of the space environment." As for building blocks, they looked for ones with a "blue-grayish hue" like cyan terracotta, stone, and light gray concrete, with yellow blocks for stripes and black blocks to mark the doors. That might seem like a small color palette for such a vast machine build, but it keeps their construction looking clean and muted, like the cutting-edge tech it's meant to represent. Especially in contrast to the deliberately chaotic meteor impact that the ship is orbiting.

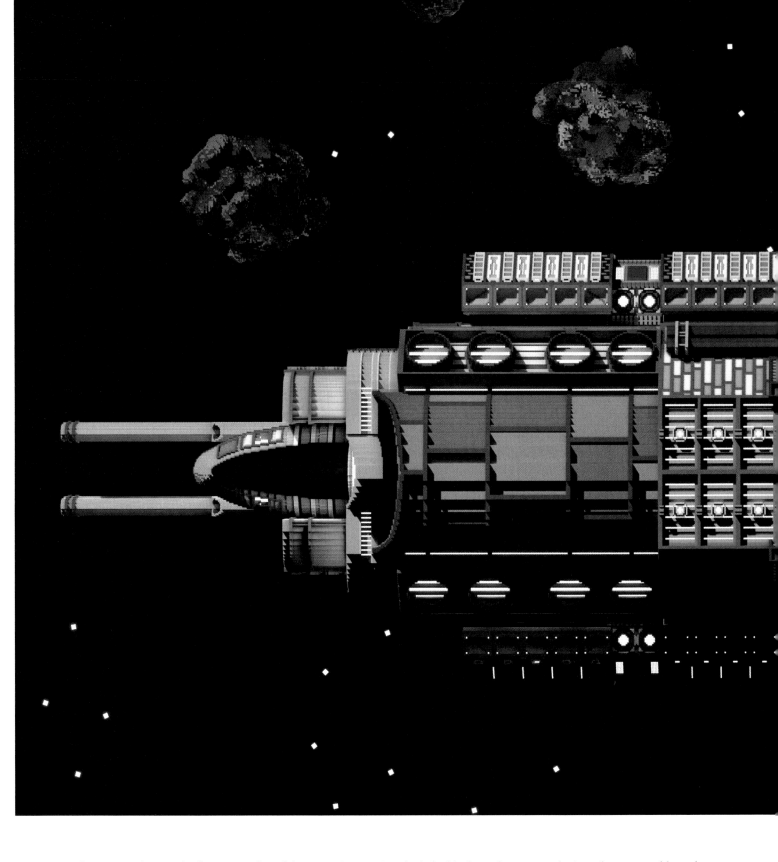

Above | The astronaut gives the scene a sense of scale.

You can see this contrast best in the flames. The fire of the meteor's impact are sharp angles of yellow and orange, striking against the black of space. Whereas the flames from the engines of the ship use blue and white fire with rounded edges, implying a controlled journey through space.

"We look for blocks with a texture that's as close as possible to the original material of the object we're trying to recreate," explains Maggie. "For the spacesuit, we used white wool, quartz – because our spacesuit reference was white. And we used stone as material for making the equipment details." NASA might not sign off on a spacesuit made of stones, but it certainly looks the part here. Once again, a small selection of colors is a deliberate choice that helps the suit resemble the real thing. The dark visor hiding the astronaut's

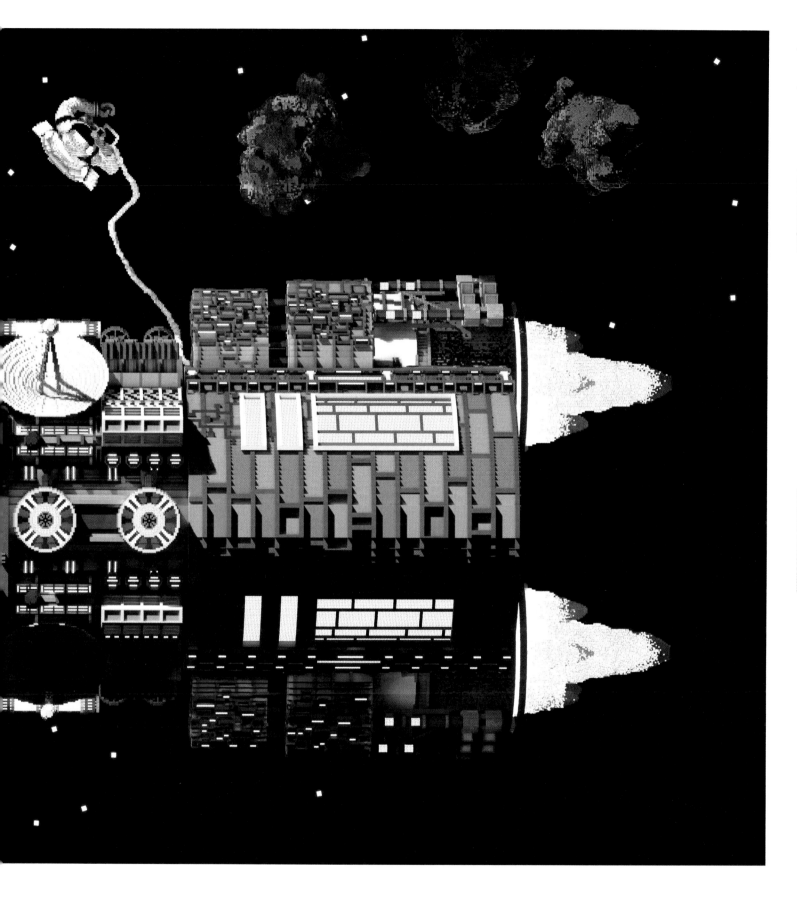

face also gives the build a sense of mystery and means they don't have to worry about a blocky recreation of a human head – always tricky to replicate in Minecraft – ruining the build's realistic vibe.

But where do you boldly go once you've checked off outer space? After discovering a passion for building through Minecraft, Maggie went on to study architecture. Having graduated, she now works for Rewrite Media, making content for Minecraft: Education Edition.

"Minecraft made me the person I am today," she says. Arkha has a similar story – he now leads MIVUBI, a team helping to develop the Minecraft community in Indonesia. "We managed to use Minecraft as a medium for presenting works of art, in collaboration with Jogja Biennale, an art foundation in Indonesia," he says. Working in teams to expand the possibilities of what Minecraft can do. Seems that defying the height limit and exploring outer space was just the beginning of this innovative duo's ambitions …

Sanctuary of the Gods & Creeperpunk

Sanctuary of the Gods

Archanee and AutomailED created this scene as a celebration of Lunar New Year. It features the four Guardians of China: the Azure Dragon, the White Leopard, the Vermillion Bird, and the Black Tortoise. In Chinese mythology, these guardians each protect one direction of the world – north, south, east, and west – but Archanee and AutomailED have placed the guardians in different areas to give the impression that they are moving.

Creeperpunk

This cyber city was designed when Archanee and AutomailED tried to imagine the future of humanity. Just what you would expect: skyscrapers, giant holograms, teleport gate, and donut shops. Once the pair had set on the design they wanted, it took them four months to finish the build. They made smart use of lights within their city, because what is a futuristic metropolis without a lot of neon? It's especially impressive when viewed at night, as the billboards and holograms give the city a dramatic vibe.

CHAPTER ELEVEN

Temple of Fortitude
By Sander Poelmans

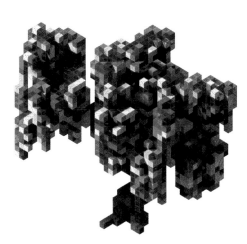

Fortitude, showing courage in adversity, is one of the most valuable traits you can have. So a build called Temple of Fortitude can't help but intrigue. Could the name refer to the temple itself? It's pretty brave to build in the middle of a body of water, after all. Perhaps it's a reference to its own builder, who showed great courage in building something so spectacular under tough circumstances? "I came up with the name just because it sounded cool, to be honest," explains Sander Poelmans, builder of the Temple of Fortitude. And he's not wrong.

Sander has an interesting planning stage for his builds – he doesn't like to place a single block on a build of this size until he's found some inspiration. "For the temple, I was heavily inspired by a great voxel artist called Mari K." It was a somewhat unfortunate source of inspiration, because the specific Mari K piece Sander was struck by contained details like blue-domed roofs and thin rectangular windows. Domes are truly one of the hardest elements to recreate in Minecraft, a game with an aforementioned love of squares and only one type of default window. Sander had no choice but to give up on his dream build. Or would that be the backstory for the Temple of Cowardice?

"After I find my inspiration, I start placing the largest shapes so I can get a feel for the general shape and composition of the build."

Sander's actual approach was to take this building challenge head on, one step at a time. "After I find my inspiration, I start placing the largest shapes so I can get a feel for the general shape and composition of the build. Once I'm happy with the shape, I'll start placing the roofs and towers. Lastly I start placing windows, trees, and other details and textures." Sander makes these steps sound simple, but this was a thorough process, which included a few more outside influences. "I didn't really research real-life temples for this, but I did make use of other references. This build uses a mix of real and made-up styles.

Above | The domed roofs on this sprawling temple are one of the core characteristics of Byzantine architecture.

Left | The different levels of the temple fit together like a huge, stylish staircase.

The towers are loosely inspired by Byzantine architecture and the flatter roofs are inspired by Mediterranean buildings. But most of the build is just made by doing what feels right!"

Sander chose to construct his build with four key colors, to provide an impactful contrast. "For the base of the building I used white concrete. Having a very plain color as a base makes it so you can more easily add color in the details! For the roofs I used red sandstone and prismarine bricks. I really like how these colors work together." But Sander felt like there was something missing in the build. "That's why I made the large pink trees in the back. I used a mix of white concrete and birch

log, and the leaves are made from all of the pink blocks in the game." Luckily, Mojang Studios included just enough different kinds of pink block – like concrete, glass, and wool – to help make those leaves pop.

Long before Sander was personifying buildings, he first encountered Minecraft: Pocket Edition, about ten years ago. "A friend and I just found it while scrolling." Absent-minded browsing led to eight years of tapping away in Pocket Edition, until Sander's ambitions outgrew this version of Minecraft. "About two years ago, I switched to Java Edition, because I wanted to do more large-scale builds. The tools and mods Java supports are just superior for that!"

He left Pocket Edition for Java because he wanted to make bigger builds, and that certainly applies to the Temple of Fortitude. The titular temple is humongous but Sander managed to speed up the process to using WorldEdit, a popular Minecraft plugin. Even the "background details" like those blossoming trees are large-scale builds in themselves. "It's quite hard to say which part I'm most pleased with, but if I had to choose one thing, it would be the large pink trees. I worked on those for at least two full days. I just really love how they look." Sander's right to be proud – curves are notoriously difficult to pull off in a crafting game that's made up of all things cube-shaped, and the way the blossoms look like they're naturally flowing off the branches is no small feat either.

Were Sander to revisit the Temple of Fortitude, he would expand the base of the building by adding more blocks. "This way it will look a bit more interesting in the large flat white areas. I would probably try to include some of the cool new blocks, like calcite, white concrete powder, and bone blocks! And add some smaller arches, indentations, and patterns!"

But these are ideas he believes builders shouldn't get hung up on. "The advice I can give for building this large is to not get stuck on small details. The most important part of builds of this size is the shape and choice of blocks you use, because those are the things you will notice the most. It's better to spend a long time on getting the shape right before trying to figure out what type of window you want to add." Sander made sure to nail the core, then decorated the structure with rich details, and celebrated by giving his temple a dramatic title.

Above | The giant pink trees are one of Sander Poelmans's favorite aspects in the Temple of Fortitude.

Left | A striking structure surrounded by water – the temple looks like it's emerging from the ocean.

CHAPTER TWELVE

Sayama City

By Sayama Production Committee

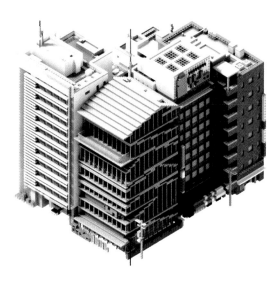

What's the longest you've spent on a Minecraft build? A week? A month? Twenty minutes before rage-quitting because a creeper blew it up? All those somewhat impressive levels of time commitment pale in the shadows of the towering skyscrapers of Sayama City, a project that – when this book comes out – will have been under construction by build team Sayama Production Committee for eight years. Eight years and counting! But when the idea is as ambitious as this – to craft an entire city, based on real cities from all over Japan – suddenly you can start to see where over ninety months of building went.

"I saw a Minecraft video on the internet and was impressed. I thought it was a great game and decided to play it," says their build team representative and founder, who prefers to remain anonymous. This was in 2012, back in Minecraft's early days, with the Sayama City project starting a year later. The original concept was to "build a Japanese city" and instead of copying one directly, they've helped themselves to influences from all over the country. Though several real places in Japan are named "Sayama," this build is not based on a real place – it's a fictional city with a lot of real-world influences.

Originally this city was the work of just one builder. But it never would have reached the heights and scale it has today if it had stayed that way. "We realized that we needed the cooperation of many people to create a large map, so we looked for people who could help us at the time we [first] published the project, which is when the team was formed." It's a clever way to put a build team together: show the community what you're working on, then recruit builders who respond to it.

Sayama Production Committee was formed, a team that "wanted to make something that no one had made before." This team has achieved exactly that. Vast cities being built by the community in Minecraft were slightly rarer in 2012, but they're a popular choice for lots of builders today, with most of the globe having been depicted in cube form.

Above | Sayama Production Committee was dedicated to building a realistic city. Small, everyday details that a real-life city planner would contemplate such as parking lots, highway bridges, and residential districts for their civilians to use.

Right | This stunning blue skyscraper towers over the rest of the skyline, certainly an architect's passion project.

Far Right | Sayama City draws from different cultures and architecture styles, including traditional Japanese temples.

"We realized that we needed the cooperation of many people to create a large map."

However, Sayama City still has most of them beat in terms of detail and sheer scope. It's the kind of build you have to zoom in on to double check you're still looking at Minecraft.

When you're building on this scale, you apparently can't afford to be picky. "We use all the blocks … every block has its own use." That could so easily have ended up a multicolored disaster with an overly busy visual palette. Instead, this team has shown admirable restraint while also using every tool the game offers. There's a lot of Minecraft's greatest gray blocks presented here, but you'd be hard pressed to call it bland. Showy displays and riots of color have taken a back seat to a commitment to verisimilitude, crafting a recreation of actual cityscapes as accurately as possible within Minecraft's famously blocky aesthetic. "In the team, each member has a specific type of building that they can build, and each of them builds according to their role."

The builders are most proud of the fact that they have stuck to that original concept. But as the years went by have they never been tempted to settle for reduced ambitions? "We've tried to make it bigger, but never tried to make it smaller," they say, giving short shrift to any suggestion of scaling down the team's ideas. This is clearly a group of builders that believe bigger is better. What could so easily have just been an impressive collection of skyscrapers actually has a nice sense of variety and clearly

● **Above** | The best word to describe Sayama Production Committee's style is realistic. They craft authentic city-living details, like pedestrian crossings and small plants growing near street lamps.

● **Right** | The Sayama City residents' shopping habits are well catered for here. Whether they want to grab lunch in the food court, buy last-minute gifts, or pick up a whole new wardrobe, there's a shop for every option.

Landmarks

For an anonymous city, Sayama is packed full of incredibly dramatic landmark structures. The Sayama Production Committee took inspiration from real-world buildings, for instance, Kyoto Tower. The team also created their own take on a skyline stand-out, such as this bold red and white Ferris wheel placed on top of an average apartment block in a residential area, like all normal fairground rides.

01 As well as roads, Sayama Production Committee has thought out high-speed public transport. The main train station sits just behind the sports center, with tracks that branch throughout the city.

02 This sports center includes a track and field arena fancy enough for the most elite of schools. They've even included floodlights.

03 The soccer pitch looks so lifelike, that's because the team made sure to use accurate scale.

04 An urban build this big could become awash with gray, but the team was careful to add some man-made greenery. A very realistic touch for the neighborhood.

05 The attention to detail within the build is stunning. Note the road markings, diligently laid down.

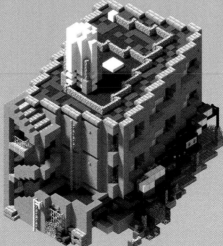

● **Below** | As cities grow, so does their population. The team has created several different residential districts filled with cute family houses.

● **Bottom** | Sayama Production Committee even imagined a kooky neighbor for their suburb. That one house whose design style stands out against the rest.

distinct districts. Larger landmarks like the Ferris wheel wouldn't be nearly as impressive or striking if they weren't against a backdrop of less extravagant buildings, carefully constructed by players determined to get all the little details right, no matter how mundane or seemingly insignificant.

That variety can likely be credited to having so many people working together on Sayama City, though there is a hierarchy when it comes to planning out the build. "Myself and other key members put together the plans," explains their representative. "There are over fifty people on the team and we communicate via Discord." The team works through a mixture of voice chat and messages, and with very few in-person

meetings, since participants come from various regions across Japan. It's something that's come up again and again when speaking to different build teams: communication is the key to a successful collaboration. Make sure you're constantly keeping each other updated and you'll see better results.

The team's plan can generally be summed up as "to express the complexity of a Japanese city, and to deform it to a size that can be created in Minecraft." Can you find an element of city life that they haven't managed to squeeze down into this build? Electronic billboards, parking spaces, office blocks, skyscrapers seemingly in every building style you'll find in Japan. That barely scratches the surface, and yet the

builders still wish the city was even wider in scope. Were they starting this build today (please don't encourage them to scrap eight years of work) they think it would be based on "cities from all over the world, not just Japan." Imagine looking upon this build and deeming it not ambitious enough, and you'll start to get an insight into the kind of mind determined enough to create a city of this scale in the first place.

Most builders, understandably, start small and work their way up. There's plenty of that here, of course, with smaller buildings helping the larger skyscrapers feel like part of a living, breathing city. But the Sayama Production Committee are clearly not builders who are capable of settling for less. "It's important to observe what you want to make and make it," they say. You might need eight years free to realize your vision, but when the results raise the bar this high for what a Minecraft city can be, what's eight years?

Left | This must be the cleanest river a large city has ever contained. The royal blue water snakes through Sayama City, providing a little glimpse of nature in a sprawling urban jungle.

Below | Helipads and mecha statues. Can you imagine anything more synonymous with a modern cityscape?

CHAPTER THIRTEEN

Rich Medieval Town

By Minecrafttalsi

TEXTURE PACK
■ **Conquest Reforged** by Conquest

A t this point, there have probably been more medieval towns built in Minecraft than there were in actual medieval times. It's a great fit, as there's always been a strong medieval fantasy aesthetic to Minecraft's world, with its wood-and-stone structures, slightly primitive tools, and terrifying dragon lurking in the darkness. No wonder so many builders have taken that theme and run with it. Builders like Minecrafttalsi, designer of this Rich Medieval Town, who likes to focus on creating small slices of medieval life, each one part of a larger fantasy world he's been working on for years. "I guess the medieval fantasy setting just fascinates me and what could be more iconic for a medieval fantasy map than a town and a castle?"

Minecrafttalsi discovered Minecraft early in 2011, when the game was still in beta. "I was browsing YouTube and found a video about a new game. I started playing the Beta 1.6 version in Survival mode, the only available mode at the time, and had an absolute blast." But Minecrafttalsi was always more interested in the building parts of Minecraft. Once Creative mode was added, he never looked back. "Fast forward ten years later, and I've worked on quite a few different maps and invested countless hours into the game. It's still a great way to channel my creativity. I probably won't quit for quite some time!"

Minecrafttalsi's got quite a build portfolio after over a decade with Minecraft, including creations as varied as a Mediterranean cemetery and a blue whale skeleton. Rather less morbidly, he's crafted many medieval builds, including estates, a monastery, and even a design for a horse and carriage. "Since the very beginning I always had the goal to create some sort of RPG world for other people to download and explore," he explains. The aim isn't just to build something easy on the eye – he wants to create a Minecraft fantasy world you can get lost in.

Not that he's at any risk of that, as he meticulously maps out his creations before starting to build properly. "I always plan out the very rough essentials first," he says of his thorough process.

● **Top and Bottom Left** | Timber-fronted houses establish the time period of the town.

● **Bottom Right** | An aerial perspective shows market stalls tucked behind the church.

Above | Minecrafttalsi found interesting details that could be added to his town, such as fountains and wooden market stalls.

Right | The baroque church that may not be historically accurate but lends itself well to the style.

"Where the town will go, where to place the castle, where to have farmland, etc. I usually get started by making a little sketch of the area. Then I mark out all the roads, places for houses, the walls, and so on. I do this in-game using different colored wool blocks." Using placeholder blocks is a smart way of getting the lay of your land before you commit to building, and there's little risk of forgetting to replace those placeholder blocks, either – players tend to notice when your "medieval church" is made of pink wool. For this particular town, Minecrafttalsi started by planning out where the main roads and gates would go. "The center of the town, where all the roads meet, had to be the marketplace and the church." Then he figured out the layout of individual alleys and houses on the fly. "For this I also often resort to manual sketches. I actually have a ton of these still in my drawer!"

Planning for Minecrafttalsi isn't just filling out a sketchbook. He's a big believer in finding reference photos and learning more about the era you're trying to recreate in Minecraft. "Doing some research before starting to build is very important. It really gives you an idea of how and why certain settlements formed." This gives him a better understanding of medieval history, but one he's not beholden to. "I personally do not build historically accurately – an obvious example being the baroque-style church – I'd rather try to create a world that is somewhat believable in and of itself. Researching actual medieval towns helps a ton with that, especially when it comes to the little details." It might seem unusual to place a baroque church in an otherwise consistently medieval town but Minecrafttalsi enjoyed the design too much not to use it here.

Above | The castle features striking timber framing and stucco walls.

Right | The wall surrounds the entire town, protecting it from possible attack.

Right Page | Inside the castle walls were decorated too, the hypothetical villagers were given picturesque courtyards, decorated with flowers.

Crossing Borders

Though the half-timbered houses are very much German in design, the locations that most influenced this build are actually the Romanian city of Râsnov and the Vosges – a range of mountains on the French-German border. Minecrafttalsi was inspired by the towns located within the Vosges. Borderline ridiculously picturesque, they're perfect fodder for rebuilding in Minecraft. *Magnifique*!

"I always use as many reference photos as possible. Mostly real-life locations or buildings, but sometimes artworks or even video games." For this town, he was mainly inspired by German half-timbered houses. These houses, with their sloped roofs and dark wooden beams that form part of the aesthetic rather than being hidden away, were common in the late medieval period, perhaps even dating back as far as the twelfth century. However, a sloping roof is a little beyond a game consisting entirely of cubes. To counteract this, Minecrafttalsi added exposed beams to his roofs to help enhance that medieval feel.

Creating a coherent town was challenging, but it was crafting the castle that proved to be the toughest part. It was an incredibly complicated design to create as his castle has "a complex and somewhat messy layout. I had to work with a lot of angled buildings." Instead of planning out a simpler castle design – that may not have suited the period setting – Minecrafttalsi took certain "artistic liberties" to make its design work in Minecraft, again taking inspiration from real-world historical styles without being beholden to them.

"I always use as many reference photos as possible. Mostly real-life locations or buildings, but sometimes artworks or even video games."

Mod-ieval Times

Minecrafttalsi strived to find the most accurate textures for his medieval build. He wanted there to be an authentic atmosphere to his town. "The map requires not only a resource pack, but a mod to run." He found a mod – Conquest Reforged for Minecraft version 1.12.2 – that also provides a wealth of blocks and textures.

"I definitely want to change it in the future. I'm actually currently working on a more realistic base map and the goal is to copy the entire town over."

That castle might have been the biggest construction challenge, but it's not his favorite element of his build. In fact, "there isn't one single part of the build that I am most proud of." Instead, he takes most pride in "how the entire place feels connected. That's personally what I like the most."

As for what he would change, Minecrafttalsi's ambitions may have resulted in him crafting himself into a corner. "Long before I built this town I created the base map. The town is located in a large valley surrounded by mountains on all sides. That turned out to be very limiting when it came to creating realistic rivers and bodies of water. All streams need to flow into the ocean. However, due to the higher elevations of all the surrounding land, I just couldn't make this work. Instead, all rivers in this area just lead to some sort of lake, which realistically would make no sense." It's a flaw that many builders would learn to live with but Minecrafttalsi has major plans to fix it. "I definitely want to change it in the future. I'm actually currently working on a more realistic base map and the goal is to copy the entire town over." Copying over an entire town just to make the rivers more realistic? With determination like that, Minecrafttalsi could probably build a castle in real life.

Painting the Town Red

Minecrafttalsi wasn't trying to make a brutal, depressingly realistic recreation of medieval times. He was aiming for a more serene fantasy land, and made sure to use colors, which reflected that. "I used a warmer color palette, which consisted mostly of earthy tones. A lot of red and brown colors especially." He even modded the biome to produce browner grass, to be more consistent with the buildings. "I think it creates a fitting atmosphere for what I was going for."

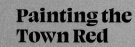

● **Top** | Creating the medieval town within the valley of a large and wooded mountain gives the build a colorful backdrop.

● **Bottom Left and Right** | The earthy tones used in the houses and surrounding vegetation provides an older, historical feel.

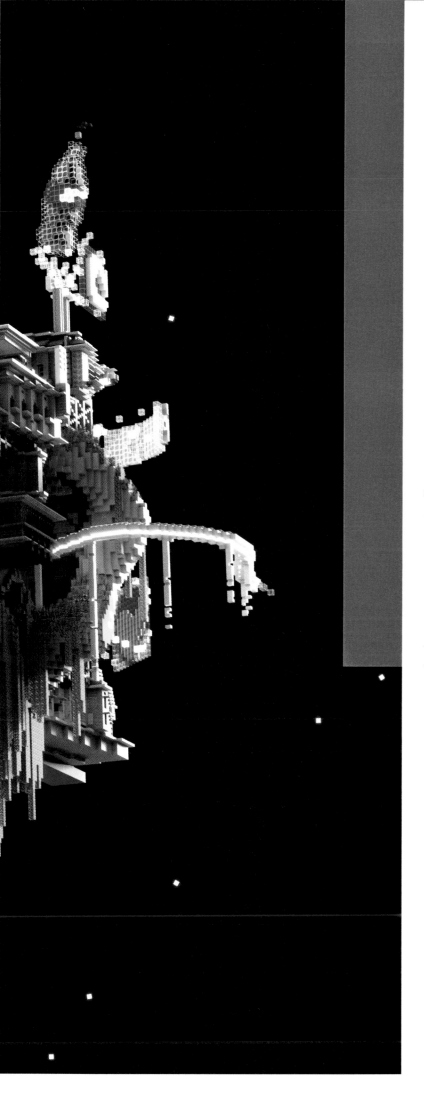

CHAPTER FOURTEEN

Dystopian Mindset

By Jossieboy

This creepy cranium-dwelling civilization named Dystopian Mindset was crafted by Jossieboy. A build where a Flat Earther would probably be considered a wishful thinker. And, like almost all Minecraft builders, he has one regret with his build. If he were to build this city in a skull today, he says he'd consider "making it scream rather than smile." Isn't this somewhat morbid metropolis plenty dark enough already? Why swap out the smile for a scream? "It just seems like that would be more powerful of an image. It might seem more painful. Which I think if you've got a city living inside your head is a fair emotion to be feeling."

Jossieboy sets out to craft things that Minecraft players have never seen before. "I really like grabbing two concepts and then smashing them together to create interesting imagery. I had this one build with an astronaut who's literally surfing through space. You take two familiar concepts and then you mash them together in interesting ways." It's a process that's inspired him in several of his other builds. He combined hermit crabs and restaurants to make a build that's somehow both, and mixed together fish and pie to make a joyously strange slice of underwater life.

"You can also, if you're lucky, end up with something that is metaphorically interesting as well. The title 'Dystopian Mindset' … you could look at it as someone being consumed by an idea. That's not how I intended it, but if you didn't know that, you very well could assume it, right? If you get the right ideas and you mash them together sometimes you strike gold."

Sometimes Jossieboy decides his themes in advance, and other times they naturally or accidentally develop as he's working on a build. When that happens, he bluffs that they were his intention all along. "You just go 'oh yeah, I meant to do that.'"

Jossieboy got into Minecraft in 2012 when he was in his early teens, long before the idea for a skull city started living in his head. "My older brother showed it to me and I was pretty much directly hooked on the idea of being able to make stuff. I didn't care much for the Survival aspect as I kept dying." But he clearly loves Creative mode. "Minecraft is basically a big ol' box of Lego and you can make whatever you want. That's what makes it so special."

Above | This eclectic city also features a giant duck statue and several digital billboards, because why not?

"I really like grabbing two concepts and then smashing them together to create interesting imagery."

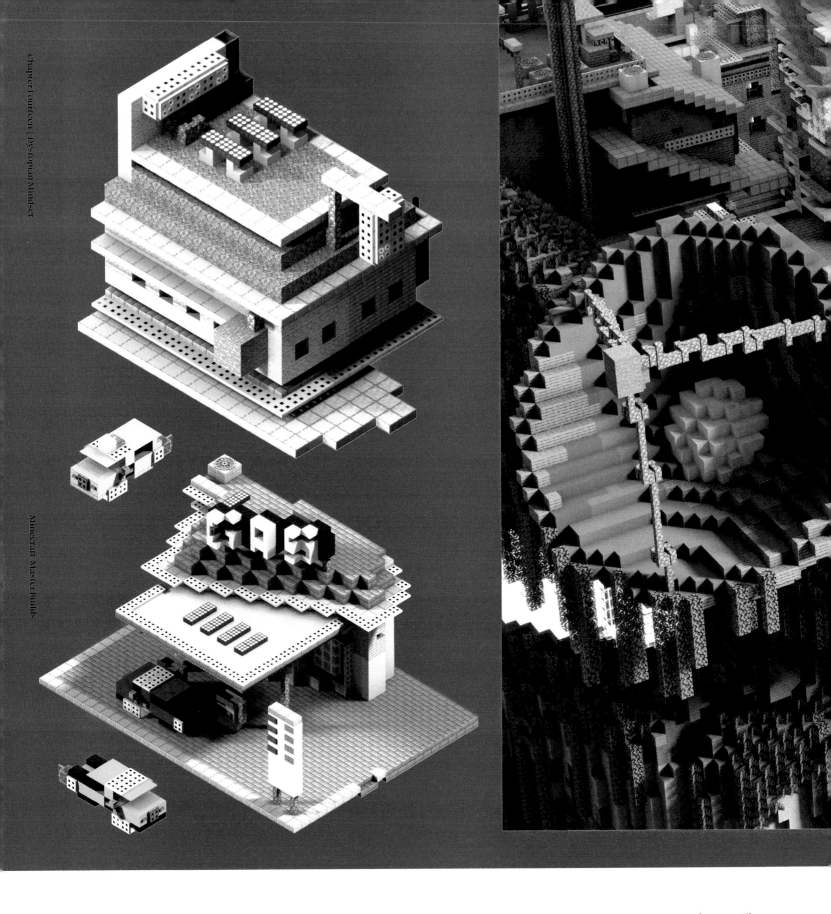

● **Above** | What mind palace is complete without multiple vehicle options and a gas station to fill them up?

● **Right** | Jossieboy used the skull as a base for the city, which means his buildings are stacked on top of each other, giving the appearance of a packed urban living space, very dystopian.

The two ideas Jossieboy combined for Dystopian Mindset were "big stone face ... but also city?" This idea was fermenting in Jossieboy's mind for some time, until he was eventually inspired to pursue it when he saw *Avengers: Infinity War.* "They visit a place called Knowhere. A mining town built into an ancient skull." Knowhere isn't just any old skeleton head – it's essentially the size of a planet. "Obviously, I don't have the scale to build that in Minecraft, that's way too massive. But the idea of a civilization living inside of an old head and mining it for resources is what I went with from that." There's even a small mining

district in Dystopian Mindset, a deliberate nod to Marvel's film. Rather than planning out a build, Jossieboy prefers to jump straight into 3D modeling. "I tend to work better in 3D. It's a lot easier for me to sculpt something with blocks. I use VoxelSniper, which allows you to create a lot of blocks really quickly. So first I shaped it, I started with the neck and jaw. That's an easy shape to make because it's a kind of arch. I looked at all kinds of references to see how a face is constructed." Jossieboy talks through how he gradually reconstructed the anatomy of a skull, block by block. "OK, I kind of have a mouth

shape and a nose shape, and that has to go up into a brow, forehead, stuff like that. I slowly shape it and then piece the brow back in. I've made the head too big, but that's OK, because I'm going to destroy most of it and then pull it back in. Add some ears and slowly I get this head shape. Then I just refine that more and more." Who knew creating life could be so complicated?

Once he'd finished the base shape of the skull, Jossieboy made a backup copy, then started the same process for where he wanted the

Shell Company

This city within a skull might not even be Jossieboy's strangest build. Another astounding construction concept he put together was "a hermit crab organic sculpture, but in the back of it it's got a ramen noodle restaurant." Because of course it does. "A little hermit crab making ramen noodles. It's got a little Zippo lighter that's heating up a bowl. Because, y'know, he's tiny. I don't know why, but the idea of taking a hermit crab and putting it in the position of being head chef of its own little restaurant that's also its shell – it just seemed adorable to me." Dystopian Mindset – although not as quirky – exhibits the same freedom of imagination.

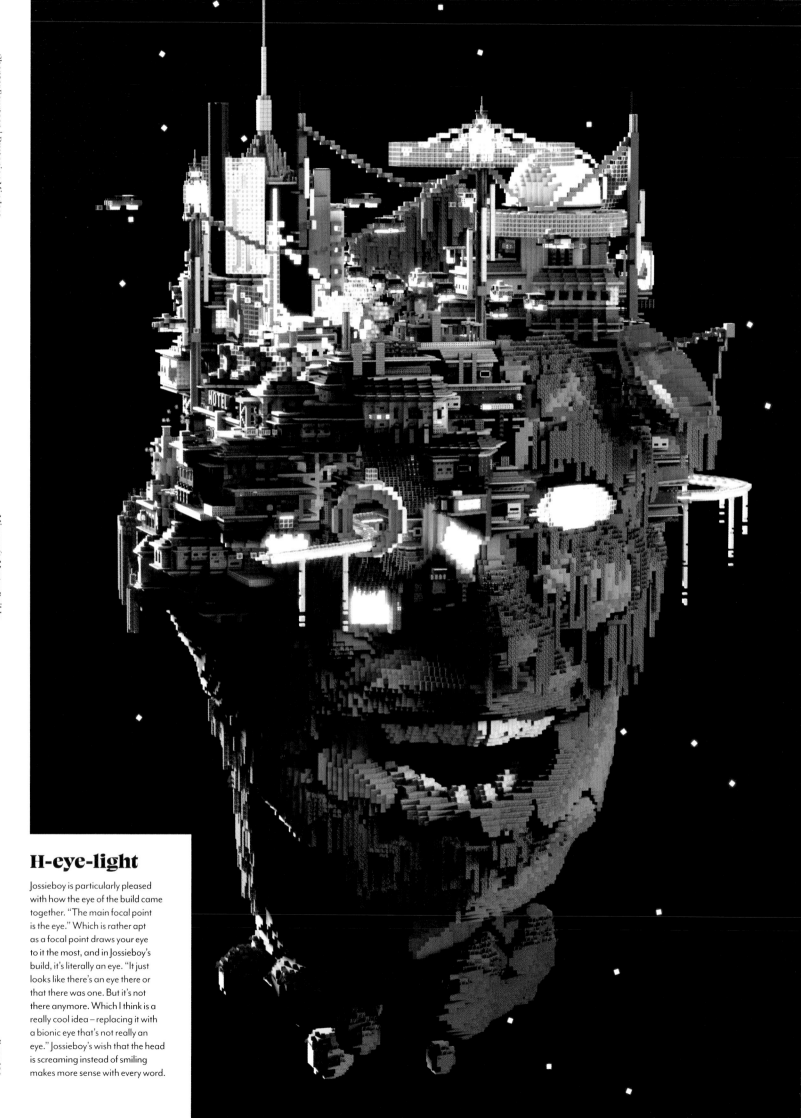

H-eye-light

Jossieboy is particularly pleased with how the eye of the build came together. "The main focal point is the eye." Which is rather apt as a focal point draws your eye to it the most, and in Jossieboy's build, it's literally an eye. "It just looks like there's an eye there or that there was one. But it's not there anymore. Which I think is a really cool idea – replacing it with a bionic eye that's not really an eye." Jossieboy's wish that the head is screaming instead of smiling makes more sense with every word.

buildings. "I wanted a very big round thing where his other eye was. I think it's really cool if you mimic the shape of architecture but it looks a lot more electronic. It's almost part of the eye, but still not. I marked that out and then started building houses on the side. Houses spreading out from his eye, almost like a tumor. I'm willing to bet there's some allegory or something there!"

Jossieboy used color-coded blocks, which he intended to later replace, for the different districts. "I know that I want contrasting stuff. If all my buildings are the same color it's not going to look good. That doesn't mean all the buildings in these blocks of color are going to be the same, just that they're going to contrast with whatever's next to it." To help that contrast, he gave the city distinct districts: the mining town, a more residential area, and a scientific research district. "That's how I went to work – base shape, color, and then details. I wanted a lot of colorful buildings to contrast with the gray and green head. I ended up using a lot of the glazed terracotta. Mostly just vibrant, colorful things to contrast the face."

A rainbow of colors was important to Jossieboy, but so was ensuring that people would immediately recognize what they were looking at, which is one of the motivations for the build's sinister smile. "The main reason he's smiling, more than anything, was because I wanted

● **Above** | Fancy staying in a hotel built within a skull city?

● **Below** | A top-down view shows how the vehicles are able to float above the city, making it easier to get to high levels.

to make it a bit creepy. But also because making closed lips would have required more shading and I wanted it to be fairly simple on the actual skull. The left side is pretty much decimated, and at one point I was planning to do that for the whole side of that face going down. But if the mouth was closed, there might not have been enough left of the face to make out that it was a face." Resulting in a build that can't help put a smile on your face, even when you know Jossieboy would be happier if it were screaming.

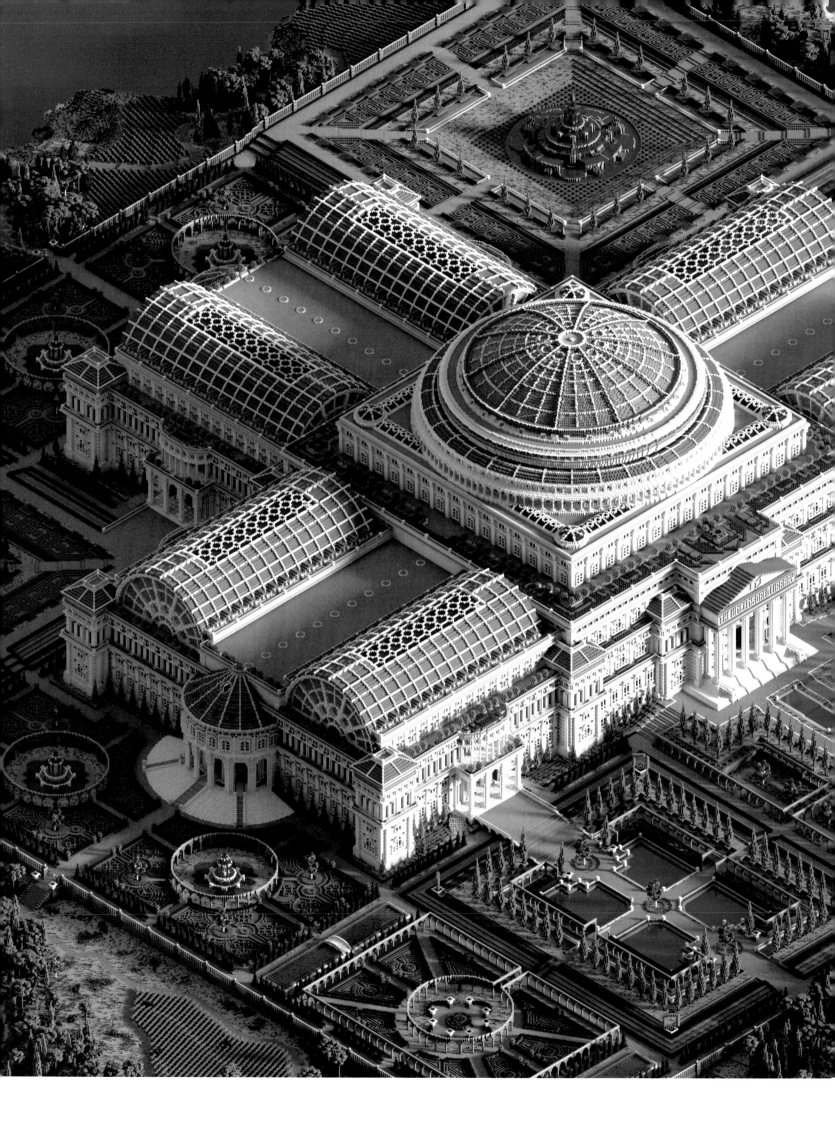

CHAPTER FIFTEEN

The Uncensored Library
By Blockworks

TEXTURE PACK
■ **Custom-made** by Blockworks

L ibraries are one of the greatest achievements of human civilization. Sanctuaries of shared knowledge, free and open to all, and a great way to enjoy this book without having to pay for it. But sadly, free information for all is far from a universal human right. It's shocking to learn how much censorship and curbs on press freedom have impacted the world. That's why Reporters Without Borders, an international nonprofit organization specializing in media freedom, collaborated with Minecraft build team Blockworks to construct The Uncensored Library. It's such an ambitious, sprawling build that "library" may even be a misnomer – Blockworks has actually crafted an entire island to explore. One that's an ingenious way of sneaking past censorship laws to get free information to everyone.

You could lose hours flying through this library's humongous halls, each room stuffed with books full of information that journalists have been persecuted, jailed, and even killed for. Blockworks has also curated several wings of the library with art installations that tell stories about how suppression of freedoms have hit parts of the world. Each one tailored to the country it is illustrating. You'll find a giant octopus in the Russia wing, for example, a "Data Kraken" with tentacles that represent the Russian government's long reach into its citizens' lives when it comes to cyber surveillance and spying.

"We tried to find countries where we could see a high degree of censorship but also where lots of people play Minecraft," explains Kristin Bässe, media and public relations officer for Reporters Without Borders. These are parts of the world that have cracked down on free expression but haven't censored Minecraft. It's a clever workaround, similar to strategies Reporters Without Borders have made use of before.

"They'd previously done a project called 'The Uncensored Playlist,'" explains James Delaney, founder and managing director of Blockworks. "Bypassing press censorship laws by sneaking censored material into song lyrics and then uploading those songs to Spotify. So countries that didn't censor Spotify but did censor this sort of content could essentially be tricked by releasing this uncensored album through Spotify."

THE UNCENSORED LIBRARY

Egypt

You'll find a towering set of scales in the Egypt wing, heavily tipped. "That's a play on the scales of justice [in ancient Egyptian culture, the god Anubis would weigh the hearts of the dead on these scales] and how weighted the justice system is in that particular country against journalists pursuing free speech. All of the rooms, the creations in them or the decoration or pattern, were designed to be attuned with that country's architectural style or history," says James Delaney.

Saudi Arabia

In the Saudi Arabia wing, you'll find an imposing cage, clashing in tone against the beauty of the majority of the library. "A lot of journalists in that particular country end up incarcerated," says James. "We wanted something that matched the grandeur of the library, but also represented the situation there. The material choice of obsidian, one of the hardest Minecraft blocks, is a deliberate choice to show how difficult it is for people to get out of those situations."

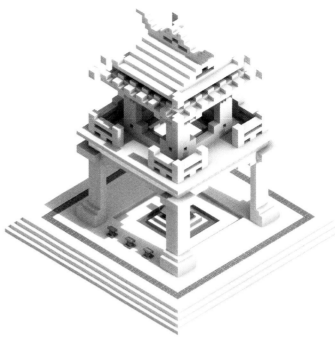

James started playing Minecraft as a teenager, meeting other players on servers and steadily building more ambitious projects. "That gradually progressed into doing it for a career. I started getting commissions from YouTubers and influencers, then Minecraft servers, then eventually Microsoft themselves, and corporations, and all sorts of people who wanted a Minecraft experience."

When Reporters Without Borders was looking for Minecraft experts who could help curate its library, Blockworks were the perfect collaborators. "It was important that it was a landmark building," James explains. "Obviously there's a lot of amazing buildings in Minecraft, so to get young kids interested, it couldn't just be a tiny library. It had to be something big and impressive." Over twelve million blocks big, eventually.

"We were looking at the Gothic libraries that you might find in universities like Oxford and Cambridge," he says, of the structures that inspired them. "We ended up with a neoclassical style, which is really the architectural style of choice for a lot of colonial powers, a lot of governments, and a lot of dictatorships as well. They gravitate toward that particular style because it asserts a certain authority."

Doesn't taking inspiration from dictatorial designs miss the point of the freedom the library is meant to represent? "It was a deliberate choice to use that style in a way that subverted the original intentions of it," James says. "Reversing the role that it's playing through the access of books inside it. Re-appropriating the neoclassical style."

Étienne-Louis Boullée, an eighteenth-century architect who was wildly overambitious for his time, was another big inspiration. "He was called 'the Paper Architect,'" says James. "He never actually built anything, because all his designs were stupidly large and splendid and magnificent. We thought that was absolutely perfect. He designed a giant cenotaph for Sir Isaac Newton, which was like an enormous

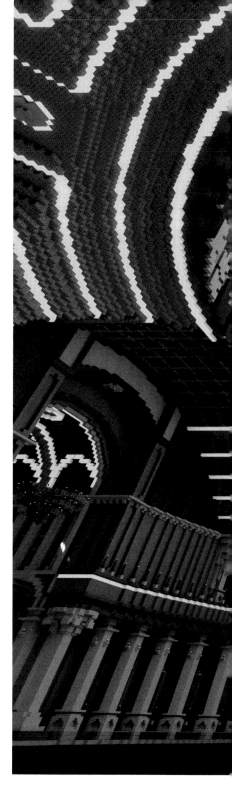

sphere. He did a potential design for the National Library of Paris, which we quite closely copied." Boullée's grand designs would be considered incredible today, let alone in the eighteenth century, when he was unfortunately far too ahead of his time for what could be achieved. Fast forward a few centuries and Blockworks can draw inspiration and finally realize his vision. "It's almost got a sci-fi feel," says James of Boullée's work. "For the 1700s, it's epic in scale and matches really well with what you can do with video games, where you don't have a budget and physics to worry about."

A bigger concern was striking the right tonal balance, given the shocking, often tragic stories the library is showcasing. "It's a difficult one because ultimately we're inside a computer game played by young people, and we're also dealing with incredibly sensitive and difficult topics," says James. "I think the Mexico room was the most challenging because there we've got the memorials of murdered journalists. Some people would say that's just not appropriate for Minecraft. But working very closely with Reporters Without Borders and making sure it's represented sensitively, talking to the families of

"It's a difficult one because ultimately we're inside a computer game played by young people."

all those involved, we felt there was a way to do it respectfully and in a way that makes people aware of the issues faced around the world."

A risky endeavor that was worth pursuing, as Minecraft helped Reporters Without Borders target a new audience. "This is how we can reach people where they actually are," says Kristin. "It's not only gamers that see the library but young people that look at YouTube videos or that find out about it from their friends. This is a very specific way to show young people that this is actually something that concerns you. It's connected to your human rights, your right to get information, to vote in a country because it's connected to democracy."

This knowledge isn't simply gained by reading in the library either. The multiplayer aspects of Minecraft helped people from different parts of

Above and Bottom Left | The grand dome takes its shape and style from neo-classical libraries across the world. Blockworks wanted the design to maintain a level of grandeur and authority.

Top Left | Representing the globe's censored works was a weighty task, and one that Reporters Without Borders and Blockworks took very seriously.

"I actually really love the statue of the fist and the pen at the beginning because this is what you see when you enter the island for the first time."

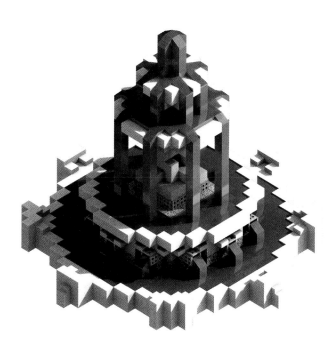

the world meet each other on the library's server. "You could connect someone from Egypt with someone from the UK in the game," explains Kristin. "The person from the UK can ask questions, like 'is it really true that this website is blocked in your country?' It's hard to imagine that they can't access a website, or they can't like a certain tweet because they would get arrested."

There are some clever perspective tricks on display. One room appears to be full of random floating blocks. However, from a certain angle, it actually spells out the word "TRUTH." "The idea with that was that a lot of the time in the countries that we're considering here, you have to really work hard to find the truth through this state-controlled media, anti-truth, and fake news," says James. It also rewards the player for engaging with the material in the library – you have to be standing next to the books to see the hidden message. "It was another way to draw the player to that location."

Reporters Without Borders has been pleased with the response to the library, and pleasantly surprised to get requests from schools and universities. "They wanted to use the library for their lectures or in their presentations or essays," Kristin tells us. "The response from gamers was really positive, and even parents! We had a lot of tweets where parents wrote things like 'I didn't know that gaming could also be so valuable!' The journalists that were involved in the campaign, I think, are happy that they can reach a different target audience than they usually do. That their texts are out there again, in this creative way that we chose to overcome censorship."

Kristin's personal favorite part of the library is how it introduces itself. "I actually really love the statue of the fist and the pen at the beginning because this is what you see when you enter the island for the first time. I think it really expresses what we want to show. But then the library slowly builds up, because usually it has to render a little bit, so it's really nice to have that moment of 'oh, where is it?'" Then steadily, block by block, a sanctuary of free information opens up. A terrific resource that's still being updated today.

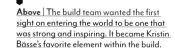

Above | The build team wanted the first sight on entering the world to be one that was strong and inspiring. It became Kristin Bässe's favorite element within the build.

CHAPTER SIXTEEN

Nova Celes City

By Quantics Build

A lot can happen in a year. You can become one year older, for example. Or, to give a slightly more ambitious example, you can construct a futuristic city in Minecraft that's gigantic in scope, has eight distinct districts, accessible interiors, and is drenched in detail to give it as much atmosphere as possible. Sounds like too much for one year? Well, that's exactly what French builder Ariux and his team Quantics Build achieved with Nova Celes. They worked relentlessly on their hyper-detailed vision of the future to meet a strict deadline. One that wasn't nearly far enough in the future.

"We finished the map on the last day," says Ariux. There's cutting it fine and then there's finishing a build on the very day you have to submit it. But that deadline couldn't be moved. Nova Celes was made for City of Swagg, an event being hosted by YouTuber Aypierre. It couldn't just be a spectacular-looking build – it had to function like a real city that Minecraft builders would be able to interact with, which meant a lot of the interiors of the city also had to be built. "It was very, very stressful because a hundred YouTubers were invited. So you couldn't push the deadline."

Those YouTubers certainly had plenty to explore. Nova Celes is made up of a central city, surrounded by eight distinct quarters. These include a Chinatown quarter, one that's a loving tribute to retro games, and even a "Biohazard" quarter. The original request had been for "just a big, big city, and we wanted to go for something set in the future because it's one of our favorite styles of build."

The build team spent six months planning out Nova Celes. That's half of their total allocated time to construct the city just preparing to build it. Not that this planning stage didn't involve any blocks – just a lot of colorful placeholders. Then six months later "the whole team was mobilized to build all the buildings and do the interiors."

"It was very, very stressful because a hundred YouTubers were invited. So you couldn't push the deadline."

Above | The centerpiece of Nova Celes is made up of a towering white structure, plenty of green spaces and utopia-style apartment blocks topped with trees.

Left | Within the outer walls of their city, Quantics Build placed boats so you can imagine how citizens would travel from quarter to quarter.

Far Left | Each quarter is guarded by a giant statue and this warrior is over looking the Chinatown district.

Tree-mendous

Knowing when to delegate is one of the most important parts of leading a build team. Ariux is a talented builder, but knew when it was best to get involved in placing blocks and when it was best to take a backseat and let his team handle construction. Luckily, delegation gets a lot easier when you're working with experts. "We had four builders who do the organic forms. They're famous organic makers in France." They also had two members who took care of the terraforming, and Ariux gives a special shout-out to the builder of the city's many trees. "We have one tree maker, very famous, he's a professional at building trees. We say 'we want a tree like this' and he makes it and it's perfect." The crafter of this fantastic foliage is French builder Cobra63, who includes in his Twitter bio "Make trees when I'm bored, so … all the time." Idle hands with naturally green fingers?

No pressure. Wait, hang on – all the pressure. With that deadline just six months away, Quantics Build crunched hard. Even the center of the build, the heart of the city that all the other quarters surround, was finished in just two weeks. "It was a huge challenge. We were fifteen people and we had to make the center, the middle part, with that big white building and lots of vegetation. It was very complicated."

Can Ariux believe he managed to pull off Nova Celes's centerpiece in a mere fourteen days? "Actually, I don't know how we made it! That was a huge challenge. I think the most complicated part to make is the ambience." Quantics Build knew from previous projects that they were skilled at building large-scale, but this time their priority was giving their future city atmosphere. It's why every street of Nova Celes is full of clever touches, little details, and the hard work of a team who've probably earned a year off building. "We think like a player when we place things on this map. It should be perfect for them because everything is for the player."

That enthusiasm bleeds into their builds – no one could look at Nova Celes and consider it the work of people who don't care for Minecraft – but sometimes reining in the team's enthusiasm is one of the most necessary evils of Ariux's role. "My favorite quarter is the Chinatown

quarter. It was our first one. We were stupid because we took a lot of time on it. We made everything. It's very dense. Lots of buildings. A lot of ambience, a lot of restaurants ... it's very, very, very huge!"

It may be his favorite quarter, but Ariux knew this approach wasn't sustainable. "We spent two months on this quarter, and I was like 'guys, we can't have this level of detail.' So for some quarters, we only had two weeks to build them. Sometimes it was hard for some quarters because we were only three or four builders. The aquatic one, I built with only two builders in two weeks." Quantics Build is made up of about fifteen to twenty builders, a number that fluctuates. "We didn't sleep a lot!"

Above | Quantics Build filled the base of the aquatic quarter with water and placed huge flat shells atop buildings.

Above Right | The base of the giant tree building demonstrates how Quantics Build created such an organic shape for their imposing tower.

Woolly World

When you're building a giant city to a tight deadline, the last thing you want to be doing is knocking down your hard work and starting again. Quantics Build sank a lot of time into planning to avoid this, but they still couldn't help making constant revisions. "For some quarters, there aren't a lot of buildings, because buildings take very long to build – two days for one person – because of all the revisions and them having to do the interior as well." One novel method they used was initially making a much softer city, building a large amount of it out of wool blocks. "We work in wool in order to have a general idea of the volumetry of the buildings. But above all, to have an idea of the density of the city. All of this helps us later to put together the city, see where we can put a park, a restaurant, a metro, etc …" Revisions are inevitable when you're working with a team of perfectionists, but consider using colorful wool as placeholder blocks when you want to see the shape of your city before you commit to its finer details!

● **Right** | Quantics Build used a lot of red blocks in their Chinatown quarter. Traditionally, the color red symbolizes luck and happiness in Chinese culture.

● **Bottom** | The team decorated the industrial quarter with smoke-topped chimneys, reminiscent of nineteenth-century factories.

● **Far Right** | The center of the build was the last quarter and featured structures with more greenery – note the two buildings with plants growing out of the walls and roof. Plus a Chinese archway – known as a *paifang* or a *pailou*.

Twenty-three-year-old Ariux founded Quantics Build in 2019, six years after he first started playing Minecraft. "At the beginning I considered us just a group of friends who played Minecraft for art." But as they took on more and more ambitious projects, they realized they were steadily becoming a professional Minecraft build team. Ariux was also studying to be an architect, fitting for such an accomplished virtual builder. Unfortunately, he was doing this while working on Nova Celes. "Recently I finished architecture school. Architecture school is very, very hard. There's a lot of work, and it was a nightmare for me to build this and do my studies."

Although both involve building, Ariux finds there are huge differences between his architectural studies and his Minecraft work. It's the unrestrained potential of the game that appealed to Ariux so much in the first place. "The problem with architecture is you have a lot of constraints. You can't do what you want. It's very rude! And in Minecraft you have the possibility to make everything." Ariux also likes the multiplayer aspect, enjoying working on a build team "because it's a melting pot of ideas. That's what I like about Minecraft, because you can do what you want with friends and the objective is to make the most impressive things."

"The problem with architecture is you have a lot of constraints. You can't do what you want. It's very rude!"

Acknowledgments

—

From the Author

First off, a massive thanks to every builder who spoke to me for this book. Your passion for Minecraft and the amazing things you've achieved with it were an absolute delight to discuss. Soon as I win the lottery, I look forward to deciding which one of you I'm going to beg to design me a house in real life. Thanks to Alex Wiltshire, who gave great advice and tracked down the majority of the builders in this book. Seriously, Alex, do you know everyone who's built something amazing in Minecraft? I barely know enough people to fill out an acknowledgments page ...

Thanks to Ellie Stores, who edited my words into coherency, answered all my dumb questions without judgment (at least I hope so?), and cut some truly awful puns that would have had you demanding I never be let near a pen again. Thanks to the designer Andy Leung and the 3D illustrator Billy Budgen, for making every page of this book look so absurdly pretty. Thanks to Jen Simpkins for being a brilliant friend. One who once told me there was some game called "Minecraft" that I should consider playing and writing about. I'll begrudgingly concede that turned out to be good career advice. Endless thanks to Adele Major, the amazing producer who collaborates with me on almost all my Minecraft projects, with the exception of this book (but she's still getting thanked in it anyway – she's that good). Finally, I'd like to thank my housemate's dog for only trying – and failing – to ruin one of my builder interviews with his incessant barking. Nice try, Odin. Better luck next book!

From the Publisher

Expanse would like to thank all the builders who constructed such impressive structures, and then took the time to talk to us about it afterward. In particular, thanks to Iskillia, SadicalMC, Thomas Sulikowski and Varuna Builds, Rigolo and Comeon, natsu3012, Rajkkor, Zeemo, qwryzu, Junghan Kim, Maggie Gondo Wardoyo, Arkha Satya Taruna, Sander Poelmans, Sayama Production Committee, Minecrafttalsi, Jossieboy, Blockworks, and Quantics Build. Thanks, also, to Kristin Bässe, LNeoX, LordKingCrown, and Craig Jelley.

A special thank you to all the people at Mojang for the guidance in bringing this book to reality. To Alex Wiltshire and Sherin Kwan, and to Cookiie and Yutaka Noma. Thank you to Mikael Hedberg for your insightful words.